Differential and Integral Calculus

VNR NEW MATHEMATICS LIBRARY

under the general editorship of

J.V. Armitage
Principal
College of St. Hild and St. Bede
Durham University

N. Curle
Professor of Applied Mathematics
University of St. Andrews

The aim of this series is to provide a reliable modern coverage of those mainstream topics that form the core of mathematical instruction in universities and comparable institutions. Each book deals concisely with a well defined key area in pure or applied mathematics or statistics. Many of the volumes are intended not solely for students of mathematics, but also for engineering and science students whose training demands a firm grounding in mathematical methods.

Titles in the series:

Differential and Integral Calculus

R.L. Wallis
Department of Engineering Design and Manufacture
University of Hull

 Van Nostrand Reinhold (UK) Co. Ltd.

Published by Van Nostrand Reinhold (UK) Co. Ltd.
Molly Millars Lane, Wokingham, Berkshire, England

Photoset in 10 on 12 pt Times
by Polyglot Pte Ltd, Singapore
Printed in Hong Kong

Library of Congress Cataloging in Publication Data

Wallis, R. L.
    Differential and integral calculus.

    (VNR new mathematics library; 11)
    Includes index.
    1. Calculus. I. Title. II. Series.
QA303.W346 1984   515   83-14506
ISBN 0-442-30578-8
ISBN 0-442-30579-6 (pbk.)

Preface

The ideas and methods of differential and integral calculus are of fundamental importance, both in the study of mathematics and in all branches of science and engineering.

In this text the ideas are introduced from basic principles. The elementary functions are introduced together with the concepts of limits, both from an intuitive standpoint and then from a more general, rigorous point of view. The text then goes on to consider the ideas of differentiation and integration, eventually leading on to the more advanced concepts of partial differentiation. Included in the development are methods for finding extreme values of functions (maxima and minima), error analysis and the ideas of Taylor series expansions for functions of one or more variables.

A text of this size cannot hope completely to exhaust the material contained in this vast subject area and this one is not an exception. The subject matter contained is that which should reasonably be included in a first-year university course for scientists and engineers. It is likely that students taking such a course will be familiar with some of the material contained in the early chapters.

I would like to record my thanks to the many people who have encouraged me during the preparation of this text and in particular to Mrs. P. Reynolds and Mrs. L.D. Whitehead for their help in typing the manuscript. Any imperfections which still remain are the sole responsibility of the author.

R.L. Wallis
University of Hull
1983

Contents

viii

1
Functions and Limits

1.1 Functions of One Variable

DEFINITION A function is a relationship that connects two sets of numbers. In this text the numbers will all be real, although the definition is still appropriate in the case of complex numbers.

If, to each member x of one set of real numbers, corresponds one or more members y of a second set of numbers then y is said to be a function of x. This functional relationship between x and y is written as

$$y = f(x) \tag{1.1.1}$$

or

$$y = y(x) \tag{1.1.2}$$

indicating symbolically that the value of y is dependent on the value of x.

Here, x is called the independent variable and y is called the dependent variable.

If to each value of x there corresponds one value of y the function is said to be single valued. If there is more than one value of y corresponding to a particular x the function is many valued.

Examples

(i) Suppose that x and y are related by the formula

$$y = x^3 \qquad \text{for all } x \tag{1.1.3}$$

then each value of x determines a unique value of y and the function is single valued.

(ii) Suppose that

$$y^2 = x \qquad \text{for } x > 0 \tag{1.1.4}$$

then for each x there are two possible values of y one positive and one negative and the function is double valued.

1

In many instances, Example (ii) being typical, the function is only defined for a certain range of values of x. In this example the function is only defined for positive values of x, since there are no real numbers y whose squares are negative.

The range of values of x for which the function is defined is called the domain of the function.

In both the examples given above the function is described by a simple formula. However in some cases, even though a perfectly good relationship between x and y exists, the relationship may not be easily described by a simple formula.

For example, suppose that a traffic census point records all the vehicles that pass that point during a particular day. If the time is denoted by x, and y denotes the number of vehicles that have passed the point up to time x, then clearly to each value of x there corresponds one value y. Hence y can be said to be a function of x, but it is improbable that a simple formula can be found to describe the function.

1.2 Graphs of Functions

Functions can be described pictorially by plotting a graph in two-dimensional cartesian space. Here the function

$$y = f(x) \qquad (1.2.1)$$

is represented by a curve, a typical point on the curve having coordinates (x, y) or $(x, f(x))$. For example the function $f(x) = x^2 - x$ is described by the curve $y = x^2 - x$ given in Figure 1.1.

In some instances the graph is the most straightforward way to examine the behaviour of a function. For example the function representing the traffic flow problem mentioned in Section 1.1 can easily be represented graphically. Any other visualization becomes extremely difficult.

1.3 Inverse Functions

DEFINITION   If the equation

$$y = f(x) \qquad (1.3.1)$$

can be solved to give x explicitly as a function of y by a formula

$$x = g(y) \qquad (1.3.2)$$

2

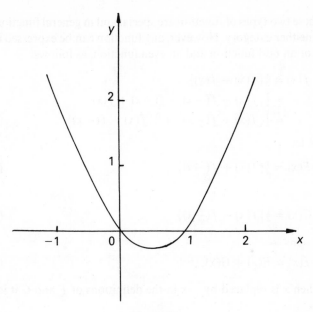

Figure 1.1

then g is the inverse of the function f. The inverse function is usually written as

$$g(x) = f^{-1}(x) \qquad (1.3.3)$$

Clearly, f is the inverse of the function g so that

$$f(x) = g^{-1}(x) \qquad (1.3.4)$$

Even in the case where $f(x)$ is single valued the inverse function $f^{-1}(x)$ may be many valued. In this instance it is usual to talk about principal values of the inverse function. This point will be amplified in Section 1.5 when some particular functions and their inverses are introduced.

1.4 Odd and Even Functions

If, when x is replaced by $-x$, the function $f(x)$ is unchanged, that is if $f(-x) = f(x)$, then $f(x)$ is called an even function. For example x^2, x^4, $x^2 + 4x^6$ are all even functions.

If, when x is replaced by $-x$, the function $f(x)$ simply changes sign, that is $f(-x) = -f(x)$, then $f(x)$ is called an odd function. For example x, x^3, $3x + 2x^5$ are all odd functions.

3

These two types of function are special and in general functions fall into neither category. However, any function can be expressed as the sum of an odd function and an even function, as follows:

$$f(x) = \tfrac{1}{2}\{f(x) + f(x)\}$$
$$= \tfrac{1}{2}\{f(x) + f(-x) - f(-x) + f(x)\}$$
$$= \tfrac{1}{2}\{f(x) + f(-x)\} + \tfrac{1}{2}\{f(x) - f(-x)\} \qquad (1.4.1)$$

Now, let

$$F(x) = \tfrac{1}{2}\{f(x) + f(-x)\} \qquad (1.4.2)$$

and

$$G(x) = \tfrac{1}{2}\{f(x) - f(-x)\} \qquad (1.4.3)$$

then

$$f(x) = F(x) + G(x) \qquad (1.4.4)$$

When x is replaced by $-x$ in the definitions of F and G it is seen that

$$F(-x) = \tfrac{1}{2}\{f(-x) + f(x)\} = F(x) \qquad (1.4.5)$$

and

$$G(-x) = \tfrac{1}{2}\{f(-x) - f(x)\} = -G(x) \qquad (1.4.6)$$

so that $F(x)$ is an even function and $G(x)$ is an odd function. Therefore $f(x)$ has been expressed, as required, as the sum of an odd function and an even function.

1.5 Some Elementary Functions

Included in this section are the definitions and simple properties of some so called elementary functions which occur in many practical problems.

Polynomials

The most elementary function is the polynomial defined as

$$f(x) = a_0 x^n + a_1 x^{n-1} + \cdots + a_{n-1} x + a_n \qquad (1.5.1)$$

where $a_0 \neq 0, a_1, a_2, \ldots, a_n$ are constants and n is a positive integer. The number n is called the degree of the polynomial.

4

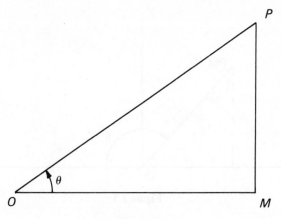

Figure 1.2

Trigonometric Functions

The trigonometric functions can be defined in various ways all of which are equivalent. For the purposes of this discussion it is sufficient to use a simple geometric definition with a slight modification for certain angle sizes.

Let θ be the angle POM in the right-angled triangle POM in Figure 1.2. Then, if θ is measured in radians, where π radians $\equiv 180°$, and $0 < \theta < \pi/2$ the elementary trigonometric functions are defined as

$$\sin\theta = \frac{PM}{OP} \qquad \cos\theta = \frac{OM}{OP} \qquad \tan\theta = \frac{PM}{OM} = \frac{\sin\theta}{\cos\theta} \qquad (1.5.2)$$

At the end points of the range the triangle degenerates into a straight line and the functions take the following values

$$\sin 0 = 0 \qquad \cos 0 = 1 \qquad\qquad\qquad (1.5.3)$$

and

$$\sin \pi/2 = 1 \qquad \cos \pi/2 = 0 \qquad\qquad\qquad (1.5.4)$$

The definition can now be extended to include angles in the range $0 \leqslant \theta \leqslant 2\pi$ (Figure 1.3).

Let P be the point (X, Y) referred to Cartesian axes OX, OY and let angle $XOP = \theta$, then for $0 \leqslant \theta \leqslant 2\pi$

$$\sin\theta = \frac{Y}{OP} = \frac{Y}{\sqrt{(X^2 + Y^2)}} \qquad \cos\theta = \frac{X}{OP} = \frac{X}{\sqrt{(X^2 + Y^2)}}$$

$$(1.5.5)$$

5

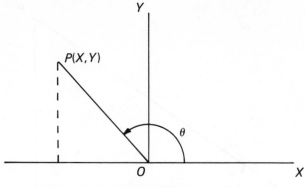

Figure 1.3

and

$$\tan\theta = \frac{Y}{X} = \frac{\sin\theta}{\cos\theta} \qquad (1.5.6)$$

Outside this range of values for θ the functions are defined to be periodic with period 2π, so that

$$\sin(2\pi + \theta) = \sin\theta \qquad \cos(2\pi + \theta) = \cos\theta$$

and

$$\tan(2\pi + \theta) = \tan\theta \qquad (1.5.7)$$

By using the periodicity of these functions the definitions can be extended to include negative angles. From Figure 1.4,

$$\sin(-\theta) = \sin(2\pi - \theta) = \frac{-Y}{R} = -\sin\theta$$

$$\cos(-\theta) = \cos(2\pi - \theta) = \frac{X}{R} = \cos\theta$$

Hence $\sin\theta$ is an odd function and $\cos\theta$ is an even function of θ.

The reciprocals of the three basic trigonometric functions are given special recognition and defined as

$$\operatorname{cosec}\theta = \frac{1}{\sin\theta} \qquad \sec\theta = \frac{1}{\cos\theta} \qquad \cot\theta = \frac{1}{\tan\theta} \qquad (1.5.8)$$

Some simple properties of these functions are given in an appendix at the end of this chapter.

6

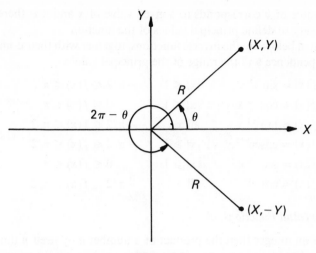

Figure 1.4

The Inverse Trigonometric Functions

If the quantities x and y are related by the formula

$$x = \sin y \tag{1.5.9}$$

then y is given by the inverse function

$$y = \sin^{-1} x \tag{1.5.10}$$

An alternative notation for this function is

$$y = \arcsin x \tag{1.5.11}$$

The remaining inverse functions

$$\cos^{-1} x \qquad \tan^{-1} x \qquad \operatorname{cosec}^{-1} x \qquad \sec^{-1} x \qquad \cot^{-1} x$$

or

$$\arccos x \qquad \arctan x \qquad \operatorname{arc\,cosec} x \qquad \operatorname{arc\,sec} x \qquad \operatorname{arc\,cot} x$$

are defined in a similar manner.

Since the trigonometric functions are periodic with period 2π, the number

$$y_1 = 2\pi + \sin^{-1} x \tag{1.5.12}$$

satisfies the condition

$$x = \sin y_1 \tag{1.5.13}$$

so that the inverse function is not single valued. That is, more than

7

one value of y corresponds to a given value of x and it is therefore necessary to define principal values of the function.

Listed below are the inverse functions, together with their domains of dependence and the range of the principal values

$$\begin{aligned}
f(x) &= \sin^{-1} x && |x| \leqslant 1 && -\pi/2 \leqslant f(x) \leqslant \pi/2 \\
f(x) &= \cos^{-1} x && |x| \leqslant 1 && 0 \leqslant f(x) \leqslant \pi \\
f(x) &= \tan^{-1} x && \text{all } x && -\pi/2 \leqslant f(x) \leqslant \pi/2 \\
f(x) &= \operatorname{cosec}^{-1} x && |x| \geqslant 1 && -\pi/2 \leqslant f(x) \leqslant \pi/2 \\
f(x) &= \sec^{-1} x && |x| \geqslant 1 && 0 \leqslant f(x) \leqslant \pi \\
f(x) &= \cot^{-1} x && \text{all } x && -\pi/2 \leqslant f(x) \leqslant \pi/2
\end{aligned}$$ (1.5.14)

Exponential Functions, a^x

If n is an integer then the product of a number a by itself n times is denoted by

$$a^n = a \cdot a \cdot a \cdots a \qquad (1.5.15)$$

where n is called the exponent.

The following operating rules apply:

$$a^{m+n} = a^m a^n \qquad a^{m-n} = a^m / a^n \qquad (a^m)^n = a^{mn} \qquad (1.5.16)$$

together with the special cases

$$a^0 = 1 \qquad \text{and} \qquad a^{-n} = 1/a^n \qquad (1.5.17)$$

This idea can now be extended to define the function

$$f(x) = a^x \qquad (1.5.18)$$

where x is a rational number, and then further to cover all real numbers.

Initially, if n is an integer, then $a^{1/n}$ is the number b satisfying the condition $b^n = a$. With this interpretation of $a^{1/n}$ the definition of a^x for rational x follows immediately.

Any rational number x can be expressed in the form

$$x = \frac{m}{n} \qquad (1.5.19)$$

where m and n are integers so that

$$a^x = a^{m/n} = (a^{1/n})^m \qquad (1.5.20)$$

The extension to irrational numbers now follows by successive approximation.

Any irrational number x can be enclosed by two rational numbers R_1 and R_2. For example, the number π lies between the rationals 3.14 and 3.145. By increasing R_1 and decreasing R_2 the irrational number can be squeezed tighter and tighter between two rationals. In the case of the number π the successive bounds may take the form

$$3.14 < \pi < 3.145 \qquad 3.141 < \pi < 3.142$$
$$3.1415 < \pi < 3.1416 \qquad \text{etc.} \tag{1.5.21}$$

At each stage values can be found for a^{R_1} and a^{R_2} using the definition (1.5.20) and as the bounds R_1 and R_2 become closer and closer so will the values of a^{R_1} and a^{R_2}. Thus by repeated approximation a limiting value can be found for $f(x) = a^x$ at the irrational point. Hence the definition for all real values x is complete.

The particular case of the above function when

$$a = e = 2.71828\ldots \tag{1.5.22}$$

is known as the exponential function and is written as

$$f(x) = e^x = \exp x \tag{1.5.23}$$

This function plays a major role in the mathematical description of many problems, exponential growth, cooling of hot bodies, radiactive decay, etc.

The significance of the number e is illustrated in appendix 1.2 of this chapter and in later chapters.

The Logarithmic Function $\log_a x$

The logarithm is the inverse function to the exponential. If x and y are related by the formula

$$x = a^y \tag{1.5.24}$$

then y can be written in terms of x as

$$y = \log_a x \tag{1.5.25}$$

The constant a is known as the base.

The logarithmic function obeys the following rules, which may be deduced directly from the operating rules for exponents

$$\log_a X + \log_a Y = \log_a(XY)$$
$$\log_a X - \log_a Y = \log_a(X/Y)$$
$$k \log_a X = \log_a(X^k) \tag{1.5.26}$$
$$\log_a 1 = 0 \qquad \text{and} \qquad \log_a a = 1$$

9

Two special cases are of particular importance. When the base $a = e$ the functions are known as the natural logarithms and are written as

$$f(x) = \log_e x = \ln x \tag{1.5.27}$$

When $a = 10$ the functions are known as Napierian logarithms which are well tabulated in many books of log tables. The Napierian logarithms are important as a tool in arithmetic calculations, but their use has been superseded by the advent of cheap electronic calculators.

Hyperbolic Functions

The hyperbolic functions are closely related to the exponential function, and as their name suggests are particularly useful in describing the geometric curve known as the hyperbola. The two basic functions are simple odd and even combinations of the positive and negative exponentials, and are defined as

$$\sinh x = \tfrac{1}{2}(e^x - e^{-x}), \qquad \cosh x = \tfrac{1}{2}(e^x + e^{-x}) \tag{1.5.28}$$

The notation used for the hyperbolic functions bears some resemblance to that used for the trigonometric functions. As was the case with the trigonometric functions, four further functions are defined as

$$\tanh x = \frac{\sinh x}{\cosh x} = \frac{e^{2x} - 1}{e^{2x} + 1} \tag{1.5.29}$$

$$\operatorname{cosech} x = \frac{1}{\sinh x} \qquad \operatorname{sech} x = \frac{1}{\cosh x} \qquad \coth x = \frac{1}{\tanh x}$$

$$\tag{1.5.30}$$

The following important properties follow immediately from the definitions:

$$\cosh x \geqslant 1$$
$$-1 \leqslant \tanh x \leqslant 1$$
$$\sinh(-x) = -\sinh x$$
$$\cosh(-x) = \cosh x$$
$$\cosh^2 x - \sinh^2 x = 1 \tag{1.5.31}$$

The hyperbolic functions have many properties which are closely related to those of the trigonometric functions but there is one major difference. The hyperbolic functions are not periodic.

Inverse Hyperbolic Functions

If x and y are related by the formula

$$x = \sinh y \qquad (1.5.32)$$

then y can be expressed in terms of x as

$$y = \sinh^{-1} x \quad \text{for all } x \qquad (1.5.33)$$

In similar fashion the five remaining inverse functions can be defined with the appropriate domains of definition as follows

$$
\begin{aligned}
&\cosh^{-1} x && x \geqslant 1 \\
&\tanh^{-1} x && |x| < 1 \\
&\operatorname{cosech}^{-1} x && \text{all } x \\
&\operatorname{sech}^{-1} x && 0 < x \leqslant 1 \\
&\coth^{-1} x && |x| > 1
\end{aligned}
\qquad (1.5.34)
$$

The three basic inverse hyperbolic functions can also be written in the form of a logarithmic function.

Suppose that $y = \sinh^{-1} x$ then $x = \sinh y$, therefore using the definition (1.5.28) it follows that

$$x = \tfrac{1}{2}(e^y - e^{-y})$$

On multiplying by $2e^y$ this equation becomes

$$2xe^y = e^{2y} - 1 \quad \text{or} \quad e^{2y} - 2xe^y - 1 = 0 \qquad (1.5.35)$$

This is a quadratic equation for e^y. This quadratic equation has solutions

$$e^y = \frac{2x \pm \sqrt{(4x^2 + 4)}}{2}$$

that is

$$e^y = x \pm \sqrt{(1 + x^2)} \qquad (1.5.36)$$

The exponential e^y must always be positive. Therefore, since $\sqrt{(x^2 + 1)} > x$, the solution containing the minus sign must be excluded to yield the single solution

$$e^y = x + \sqrt{(x^2 + 1)} \qquad (1.5.37)$$

This exponential can now be inverted to give

$$y = \ln[x + \sqrt{(x^2 + 1)}]$$

11

Hence

$$\sinh^{-1} x = \ln[x + \sqrt{(x^2 + 1)}] \tag{1.5.38}$$

Similarly using the definitions for $\cosh^{-1} x$ and $\tanh^{-1} x$ it can be shown that

$$\cosh^{-1} x = \pm \ln[x + \sqrt{(x^2 - 1)}] \tag{1.5.39}$$

and

$$\tanh^{-1} x = \tfrac{1}{2} \ln \frac{1 + x}{1 - x} \tag{1.5.40}$$

1.6 Limits

In the introduction to the exponential function a^x the idea of the value of a function at a point being obtained as the limit of successive approximations was used intuitively. Here, these ideas are extended and placed on a firm foundation.

Consider the function

$$f(x) = 3 + \frac{1}{x^2} \tag{1.6.1}$$

As x increases the function decreases and as x becomes infinitely large $f(x)$ approaches the value 3. There is no value of x for which $f(x)$ is actually equal to the value 3, but by choosing x sufficiently large the value of $f(x)$ can be made as close to 3 as desired.

Here $f(x)$ is said to tend to the limit 3 as x tends to infinity.

In this example the limiting behaviour is described when x becomes large. Similar limiting processes occur at finite points. Consider the function

$$f(x) = \frac{\sin x}{x} \tag{1.6.2}$$

This function is well behaved everywhere except when $x = 0$. At this point, both x and $\sin x$ vanish and the function is undefined. However from Table 1.1 it can be seen that for small values of x, measured in radians, the values of x and $\sin x$ are almost the same and the ratio $\sin x / x$ is almost unity. As x becomes closer to zero, the ratio becomes closer and closer to the value 1 and it can be said that in the limit as $x = 0$ the function $f(x)$ has limit 1.

Table 1.1

x in degrees	1	2	3	4	5	6
x in radians	0.0174533	0.0349066	0.0523599	0.0698132	0.0872665	0.1047198
sin x	0.0174524	0.0348995	0.052336	0.0697565	0.0871557	0.1045285
sin x/x	0.9999484	0.9997966	0.9995435	0.9991878	0.9987303	0.9981729

Again there is no value of x for which $f(x) = 1$, but by taking x sufficiently small, the value of $f(x)$ can be made as close to 1 as one likes. A formal proof of this limit is given in Appendix 1.2.

These ideas of limiting processes can now be formalized in the form of a definition.

DEFINITION   If $f(x)$ is a function, defined for all values of x, except possibly $x = a$, and for all values of x sufficiently close to $x = a$ the difference between $f(x)$ and a constant value L can be made arbitrarily small, then $f(x)$ is said to tend to limit L as x approaches the value $x = a$.

This limit is written as

$$\lim_{x \to a} f(x) = L \qquad (1.6.3)$$

The above definition can be expressed symbolically as follows.

Given an arbitrarily small quantity $\varepsilon > 0$, if there exists a positive value δ (usually dependent on ε) such that

$$|f(x) - L| < \varepsilon \qquad (1.6.4)$$

whenever x satisfies the condition

$$|x - a| < \delta \qquad (1.6.5)$$

then

$$\lim_{x \to a} f(x) = L$$

Example   Show formally that

$$\lim_{x \to 0} \left[\frac{\sqrt{(1 + x^2)} - 1}{x} \right] = 0$$

Let $\varepsilon > 0$, be given, then it is necessary to show that there exists a number δ such that

$$\left| \frac{\sqrt{(1 + x^2)} - 1}{x} \right| < \varepsilon$$

for all x such that $|x| < \delta$.
If $x > 0$

$$\left| \frac{\sqrt{(1 + x^2)} - 1}{x} \right| < \varepsilon$$

gives

$$\frac{\sqrt{(1 + x^2)} - 1}{x} < \varepsilon$$

so that

$$\sqrt{(1 + x^2)} < 1 + \varepsilon x \qquad (1.6.6)$$

Hence, squaring both sides of the inequality,

$$1 + x^2 < 1 + 2\varepsilon x + \varepsilon^2 x^2$$

giving

$$(1 - \varepsilon^2)x^2 < 2\varepsilon x \qquad (1.6.7)$$

Now $x > 0$, therefore division of both sides by x gives

$$(1 - \varepsilon^2)x < 2\varepsilon$$

that is

$$x < \frac{2\varepsilon}{1 - \varepsilon^2} \qquad (1.6.8)$$

If $x < 0$

$$\left| \frac{\sqrt{(1 + x^2)} - 1}{x} \right| < \varepsilon$$

gives

$$\frac{\sqrt{(1 + x^2)} - 1}{-x} < \varepsilon$$

so that

$$\sqrt{(1 + x^2)} < 1 - \varepsilon x \qquad (1.6.9)$$

Again squaring both sides of the inequality

$$1 + x^2 < 1 - 2\varepsilon x + \varepsilon^2 x^2$$

giving

$$(1 - \varepsilon^2)x^2 < -2\varepsilon x \qquad (1.6.10)$$

Now $x < 0$, therefore division by x gives

$$(1 - \varepsilon^2)x > -2\varepsilon$$

that is

$$x > \frac{-2\varepsilon}{1 - \varepsilon^2} \qquad (1.6.11)$$

Hence, provided that $-2\varepsilon/(1 - \varepsilon^2) < x < 2\varepsilon/(1 - \varepsilon^2)$ it follows that

$$\left| \frac{\sqrt{(1 + x^2)} - 1}{x} \right| < \varepsilon$$

In this example $\delta = 2\varepsilon/(1 - \varepsilon^2)$ and the existance of the limit is formally established.

In many cases, as x approaches a given value, no finite limit exists, but the function either increases or decreases indefinitely without limit.

For example the function $f(x) = 1/(x - a)^2$ increases without limit as x approaches the value $x = a$.

In such cases $f(x)$ is said to tend to either plus infinity or minus infinity, depending whether the function increases or decreases, and the limit is written as

$$\lim_{x \to a} f(x) = \infty \qquad \text{or} \qquad \lim_{x \to a} f(x) = -\infty \qquad (1.6.12)$$

The above definition places no restrictions on the way in which the value $x = a$ is approached. It is however important, with some functions to examine separately the limit as $x = a$ is approached through values of x that are greater than a and the limit when $x = a$ is approached through values less than a.

These two limits are written as

$$f(a^+) = \lim_{x \to a^+} f(x) \qquad \text{and} \qquad f(a^-) = \lim_{x \to a^-} f(x) \qquad (1.6.13)$$

where the notation $x \to a^+$ indicates that the value $x = a$ is approached from above and $x \to a^-$ indicates that the value $x = a$ is approached from below.

Sometimes these limits are called right-hand and left-hand limits, respectively.

The limit $\lim_{x \to a} f(x) = L$ exists if, and only if, $f(a^+) = f(a^-)$ then

$$\lim_{x \to a} f(x) = f(a^+) = f(a^-) \qquad (1.6.14)$$

The simplest case of a limit occurs when the function is defined at $x = a$ and $\lim_{x \to a} f(x)$ exists and is equal to $f(a)$, the value of the function at $x = a$. In this case the function is said to be continuous at the point $x = a$.

16

As was the case with the first example in this section it is frequently necessary to examine the behaviour of a function as x increases or decreases without bound, that is as $x \to \infty$ or as $x \to -\infty$, respectively. In this case the definition of the limit has to be modified.

DEFINITION   If given any number $\varepsilon > 0$, no matter how small, there exists a positive number N (usually dependent on ε) such that

$$|f(x) - L| < \varepsilon \qquad (1.6.15)$$

whenever $x > N$ then $f(x)$ is said to tend to L as x tends to infinity.
    The limit is written as

$$\lim_{x \to \infty} f(x) = L \qquad (1.6.16)$$

or the behaviour is described by writing

$$f(x) \to L \quad \text{as} \quad x \to \infty \qquad (1.6.17)$$

For example the limit obtained for $f(x) = 3 + 1/x^2$, earlier, can now be demonstrated formally.
    Given any $\varepsilon > 0$

$$|f(x) - 3| = \left| 3 + \frac{1}{x^2} - 3 \right|$$

$$= \frac{1}{x^2}$$

$$< \varepsilon$$

provided that

$$\frac{1}{x^2} < \varepsilon$$

Hence $|f(x) - 3| < \varepsilon$ whenever $x > 1/\varepsilon^{1/2}$, so that in this example $N = 1/\varepsilon^{1/2}$.

1.7   Properties of Limits

If $f(x)$ and $g(x)$ are two functions of x and the limits $\lim_{x \to a} f(x) = F$ and $\lim_{x \to a} g(x) = G$ both exist, then

(i) $\displaystyle \lim_{x \to a} \{f(x) \pm g(x)\} = F \pm G$ $\qquad (1.7.1)$

17

(ii) $\displaystyle\lim_{x \to a} f(x)g(x) = F \cdot G$ (1.7.2)

(iii) $\displaystyle\lim_{x \to a} f(x)/g(x) = F/G$, provided that $G \neq 0$ (1.7.3)

Appendix 1.1   Combination Formulae for Trigonometric and Hyperbolic Functions

Let the angles θ and ϕ be as shown in Figure 1.5, then it follows from the definition of the trigonometric function (1.5.2), using right-angled triangle OPC, that

$$
\begin{aligned}
\sin(\theta + \phi) &= \frac{PC}{OP} \\[2mm]
&= \frac{PA + AC}{OP} \\[2mm]
&= \frac{PA + BD}{OP} \\[2mm]
&= \frac{PA}{PB} \cdot \frac{PB}{OP} + \frac{BD}{OB} \cdot \frac{OB}{OP} \\[2mm]
&= \cos\theta \sin\phi + \sin\theta \cos\phi
\end{aligned}
$$

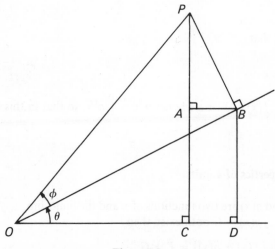

Figure 1.5

18

Hence
$$\sin(\theta + \phi) = \sin\theta\cos\phi + \cos\theta\sin\phi \qquad (A1.1)$$
Similarly,
$$\cos(\theta + \phi) = \frac{OC}{OP}$$
$$= \frac{OD - CD}{OP}$$
$$= \frac{OD - AB}{OP}$$
$$= \frac{OD}{OB} \cdot \frac{OB}{OP} - \frac{AB}{PB} \cdot \frac{PB}{OP}$$
Hence
$$\cos(\theta + \phi) = \cos\theta\cos\phi - \sin\theta\sin\phi \qquad (A1.2)$$
If ϕ is replaced by $-\phi$ these two combinations become
$$\sin(\theta - \phi) = \sin\theta\cos\phi -- \cos\theta\sin\phi \qquad (A1.3)$$
and
$$\cos(\theta - \phi) = \cos\theta\cos\phi + \sin\theta\sin\phi \qquad (A1.4)$$
By definition
$$\tan(\theta + \phi) = \frac{\sin(\theta + \phi)}{\cos(\theta + \phi)}$$
$$= \frac{\sin\theta\cos\phi + \cos\theta\sin\phi}{\cos\theta\cos\phi - \sin\theta\sin\phi}$$
Therefore, dividing the numerator and denominator by $\cos\theta\cos\phi$, it follows that
$$\tan(\theta + \phi) = \frac{\tan\theta + \tan\phi}{1 - \tan\theta\tan\phi} \qquad (A1.5)$$
Similarly,
$$\tan(\theta - \phi) = \frac{\tan\theta - \tan\phi}{1 + \tan\theta\tan\phi} \qquad (A1.6)$$
Special cases of these results follow by setting $\phi = \theta$. Hence
$$\cos^2\theta + \sin^2\theta = 1 \qquad (A1.7)$$

19

$$\sin 2\theta = 2 \sin \theta \cos \theta \qquad \text{(A1.8)}$$

$$\cos 2\theta = \cos^2 \theta - \sin^2 \theta \qquad \text{(A1.9)}$$

$$= 2 \cos^2 \theta - 1 \qquad \text{(A1.10)}$$

$$= 1 - 2 \sin^2 \theta \qquad \text{(A1.11)}$$

and

$$\tan 2\theta = \frac{2 \tan \theta}{1 - \tan^2 \theta} \qquad \text{(A1.12)}$$

Division of (A1.7) in turn by $\cos^2 \theta$ and $\sin^2 \theta$ gives

$$\sec^2 \theta = 1 + \tan^2 \theta \qquad \text{(A1.13)}$$

and

$$\operatorname{cosec}^2 \theta = 1 + \cot^2 \theta \qquad \text{(A1.14)}$$

respectively.

Equations (A1.1) and (A1.3) can be combined to give

$$\sin(\theta + \phi) + \sin(\theta - \phi) = 2 \sin \theta \cos \phi$$

and

$$\sin(\theta + \phi) - \sin(\theta - \phi) = 2 \cos \theta \sin \phi$$

Similarly using (A1.2) and (A1.4)

$$\cos(\theta + \phi) + \cos(\theta - \phi) = 2 \cos \theta \cos \phi$$

and

$$\cos(\theta + \phi) - \cos(\theta - \phi) = -2 \sin \theta \sin \phi$$

In these expressions let $\theta + \phi = \alpha$ and $\theta - \phi = \beta$ so that

$$\theta = \tfrac{1}{2}(\alpha + \beta) \quad \text{and} \quad \phi = \tfrac{1}{2}(\alpha - \beta)$$

It then follows that

$$\sin \alpha + \sin \beta = 2 \sin \frac{\alpha + \beta}{2} \cos \frac{\alpha - \beta}{2} \qquad \text{(A1.15)}$$

$$\sin \alpha - \sin \beta = 2 \cos \frac{\alpha + \beta}{2} \sin \frac{\alpha - \beta}{2} \qquad \text{(A1.16)}$$

$$\cos \alpha + \cos \beta = 2 \cos \frac{\alpha + \beta}{2} \cos \frac{\alpha - \beta}{2} \qquad \text{(A1.17)}$$

$$\cos \alpha - \cos \beta = -2 \sin \frac{\alpha + \beta}{2} \sin \frac{\alpha - \beta}{2} \qquad \text{(A1.18)}$$

Some useful values of the trigonometric functions are given in Table 1.2.

Table 1.2

θ	0	$\pi/6$	$\pi/4$	$\pi/3$	$\pi/2$
$\sin\theta$	0	$1/2$	$1/\sqrt{2}$	$\sqrt{3}/2$	1
$\cos\theta$	1	$\sqrt{3}/2$	$1/\sqrt{2}$	$1/2$	0
$\tan\theta$	0	$1/\sqrt{3}$	1	$\sqrt{3}$	∞

The hyperbolic functions satisfy similar properties to the trigonometric functions with slight modifications. Those properties which follow directly from the definitions are listed below.

$$\cosh^2 x - \sinh^2 x = 1 \tag{A1.19}$$

$$1 - \tanh^2 x = \operatorname{sech}^2 x \tag{A1.20}$$

$$\coth^2 x - 1 = \operatorname{cosech}^2 x \tag{A1.21}$$

$$\sinh(x \pm y) = \sinh x \cosh y \pm \cosh x \sinh y \tag{A1.22}$$

$$\cosh(x \pm y) = \cosh x \cosh y \pm \sinh x \sinh y \tag{A1.23}$$

$$\tanh(x \pm y) = \frac{\tanh x \pm \tanh y}{1 \pm \tanh x \tanh y} \tag{A1.24}$$

$$\sinh x + \sinh y = 2\sinh\frac{x+y}{2}\cosh\frac{x-y}{2} \tag{A1.25}$$

$$\sinh x - \sinh y = 2\cosh\frac{x+y}{2}\sinh\frac{x-y}{2} \tag{A1.26}$$

$$\cosh x + \cosh y = 2\cosh\frac{x+y}{2}\cosh\frac{x-y}{2} \tag{A1.27}$$

$$\cosh x - \cosh y = 2\sinh\frac{x+y}{2}\sinh\frac{x-y}{2} \tag{A1.28}$$

$$\sinh 2x = 2\sinh x \cosh x \tag{A1.29}$$

$$\cosh 2x = \cosh^2 x + \sinh^2 x \tag{A1.30}$$

Appendix 1.2   Evaluation of Two Important Limits

(i) $\displaystyle\lim_{\theta\to 0}\frac{\theta}{\sin\theta} = 1$ (Figure 1.6)

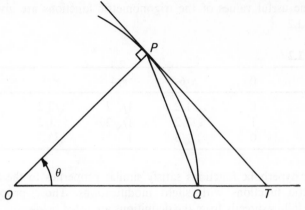

Figure 1.6

Let points P and Q lie on the circumference of a circle centre O and let the tangent to the circle at the point P meet the extension of line OQ at the point T.

It is evident from the geometry that

Area of triangle POQ < Area of sector POQ < Area of triangle POT.

Therefore, if θ is the angle POQ measured in radians,

$$\tfrac{1}{2}OP^2 \sin \theta < \tfrac{1}{2}OP^2 \theta < \tfrac{1}{2}OP^2 \tan \theta$$

Division of these inequalities by $\tfrac{1}{2}OP^2 \sin \theta$ gives

$$1 < \frac{\theta}{\sin \theta} < \frac{1}{\cos \theta}$$

Hence in the limit as θ approaches zero,

$$1 \leqslant \lim_{\theta \to 0} \frac{\theta}{\sin \theta} \leqslant \lim_{\theta \to 0} \frac{1}{\cos \theta}$$

Now, $\lim_{\theta \to 0} \cos \theta = 1$, so the inequality becomes

$$1 \leqslant \lim_{\theta \to 0} \frac{\theta}{\sin \theta} \leqslant 1$$

If both conditions are to be satisfied the equalities must hold, therefore

$$\lim_{\theta \to 0} \frac{\theta}{\sin \theta} = 1 \tag{A2.1}$$

22

This limit may also be written as

$$\lim_{\theta \to 0} \frac{\sin \theta}{\theta} = 1 \qquad \text{(A2.2)}$$

(ii) $\lim_{x \to \infty} \left(1 + \dfrac{1}{x}\right)^x = e$

Let $a_n = (1 + 1/n)^n$, where n is a positive integer, be a sequence of numbers.

By the use of the Binomial theorem, a_n can be expanded as

$$a_n = \left(1 + \frac{1}{n}\right)^n = \underline{n} + n \cdot \frac{1}{n} + \frac{n(n-1)}{2!}\frac{1}{n^2} + \frac{n(n-1)(n-2)}{3!}\frac{1}{n^3} + \cdots$$

$$+ \frac{n(n-1)(n-2)\ldots 1}{n!} \cdot \frac{1}{n^n}$$

$$= 1 + 1 + \frac{1}{2!}\left(1 - \frac{1}{n}\right) + \frac{1}{3!}\left(1 - \frac{1}{n}\right)\left(1 - \frac{2}{n}\right) + \cdots$$

$$+ \frac{1}{n!}\left(1 - \frac{1}{n}\right)\left(1 - \frac{2}{n}\right)\cdots\left(1 - \frac{n-1}{n}\right)$$

When n is replaced by $n + 1$ in this expansion it is clear that each term in the expansion for a_{n+1} is at least as large as the corresponding term in the expansion for a_n. In addition a_{n+1} contains an extra positive term.
Hence

$$a_n < a_{n+1}$$

therefore the sequence a_n increases as n increases.

Since a_n is an increasing sequence it must either tend to a finite limit or become infinite as n tends to infinity.

From the expansion of a_n it follows that

$$a_n < 1 + 1 + \frac{1}{1 \cdot 2} + \frac{1}{1 \cdot 2 \cdot 3} + \cdots + \frac{1}{1 \cdot 2 \cdot 3 \cdots n}$$

$$< 1 + 1 + \frac{1}{2} + \frac{1}{4} + \cdots + \frac{1}{2^{n-1}}$$

$$< 3$$

Again, from the expansion, since every term is positive the value of a_n

23

must be greater than the sum of the first two terms, that is

$$a_n > 1 + 1 = 2$$

Combining both these inequalities

$$2 < a_n < 3$$

so that a_n is contained between finite limits and must therefore tend to a finite limit as n tends to infinity. This limit is defined to be the number e.

Hence, letting n tend to infinity

$$\lim_{n \to \infty} \left(1 + \frac{1}{n} \right)^n = e \qquad \text{(A2.3)}$$

where $2 < e < 3$.

The value of the number e can be calculated to any required degree of accuracy, and it can be shown that, correct to six significant figures

$$e = 2.71828 \qquad \text{(A2.4)}$$

The above result can now be extended to find the limit, as $x \to \infty$, of the function

$$f(x) = \left(1 + \frac{1}{x} \right)^x$$

where x can assume, not only integer, but all real values.

Let n be the largest integer which is less than x, then

$$1 + \frac{1}{x} < 1 + \frac{1}{n} \qquad \text{and} \qquad x < n + 1$$

Consequently

$$\left(1 + \frac{1}{x} \right)^x < \left(1 + \frac{1}{n} \right)^{n+1} \qquad \text{(A2.5)}$$

Similarly it follows that

$$\left(1 + \frac{1}{n+1} \right)^n < \left(1 + \frac{1}{x} \right)^x \qquad \text{(A2.6)}$$

Hence

$$\left(1 + \frac{1}{n+1} \right)^n < \left(1 + \frac{1}{x} \right)^x < \left(1 + \frac{1}{n} \right)^{n+1} \qquad \text{(A2.7)}$$

It has already been shown that

$$\lim_{n \to \infty} \left(1 + \frac{1}{n}\right)^n = e$$

therefore

$$\lim_{n \to \infty} \left(1 + \frac{1}{n}\right)^{n+1} = \lim_{n \to \infty} \left(1 + \frac{1}{n}\right) \cdot \lim_{n \to \infty} \left(1 + \frac{1}{n}\right)^n$$

$$= 1 \cdot e$$

$$= e$$

Also

$$\lim_{n \to \infty} \left(1 + \frac{1}{n+1}\right)^n = \lim_{n \to \infty} \frac{1}{\left(1 + \dfrac{1}{n+1}\right)} \cdot \lim_{n \to \infty} \left(1 + \frac{1}{n+1}\right)^{n+1}$$

$$= 1 \cdot e$$

$$= e$$

Hence, in the limit as x tends to infinity, the inequality (A2.7) gives

$$e \leqslant \lim_{x \to \infty} \left(1 + \frac{1}{x}\right)^x \leqslant e$$

For both conditions to be satisfied the equality must hold, therefore

$$\lim_{x \to \infty} \left(1 + \frac{1}{x}\right)^x = e$$

Exercises

1.1 For which values of x are each of the following functions defined?

(i) $(1 + x)^{1/2}$ (ii) $(1 + x^2)^{1/2}$

(iii) $\sin^{-1}(1 - x)$ (iv) $\ln x$

(v) $\tanh^{-1}(1 - x^2)$ (vi) $\cos^{-1}(1 + x^2)$

(vii) $\cosh^{-1}(1 + x)$ (viii) $\ln(\sinh x)$

1.2 State whether the following functions are odd or even.

(i) $\sin x^2$ (ii) $x + \sin x$

(iii) $\tan x$ (iv) $x \tanh x$

(v) $\sec x$ (vi) $\ln[(1 - x)/(1 + x)]$

(vii) $x^3 \ln[(1 - x)/(1 + x)]$ (viii) $(x + 1/x)^2$

13. $$\frac{1 + 2\sin\frac{x}{2}\cos\frac{x}{2} + 2\sin^2\frac{x}{2} - 1}{1 + 2\sin\frac{x}{2}\cos\frac{x}{2} + 2\cos^2\frac{x}{2} - 1} = \frac{\sin\frac{x}{2}\cos\frac{x}{2} + \sin^2\frac{x}{2}}{\sin\frac{x}{2}\cos\frac{x}{2} + \cos^2\frac{x}{2}}$$

$$= \frac{\tan\frac{x}{2} + \tan^2\frac{x}{2}}{\tan\frac{x}{2} + 1} = \tan\frac{x}{2}$$

1.3 Show that

$$\frac{1 + \sin x - \cos x}{1 + \sin x + \cos x} = \tan\frac{1}{2}x$$

and

$$\frac{\sin x + \sin 3x + \sin 5x}{\cos x + \cos 3x + \cos 5x} = \tan 3x$$

1.4 If

$$\cos x + \cos 3x = \cos 2x + \cos 4x$$

show that either $\cos x = 0$ or $\cos 3x = \cos 2x$. Hence find all the values of x for which the original equation is satisfied.

1.5 Prove the formula

$$\cosh(a + b) = \sinh a \sinh b + \sqrt{[(1 + \sinh^2 a)(1 + \sinh^2 b)]}$$

1.6 Prove the following
   (i) $\log_a x = \log_a b \cdot \log_b x$
   (ii) $\log_a b \cdot \log_b c \cdot \log_c a = 1$

   (iii) $\dfrac{1}{\log_a(abc)} + \dfrac{1}{\log_b(abc)} + \dfrac{1}{\log_c(abc)} = 1$

1.7 Prove the formula

$$\tan^{-1} A - \tan^{-1} B = \tan^{-1}\left(\frac{A - B}{1 + AB}\right)$$

What is the corresponding result for $\tanh^{-1} A - \tanh^{-1} B$?

1.8 Find the following limits

   (i) $\displaystyle\lim_{x\to 1}\left(\frac{x - 1}{x^2 - 1}\right)$ (ii) $\displaystyle\lim_{x\to 0}\left(\frac{x\sin x}{x^2 + x}\right)$

   (iii) $\displaystyle\lim_{x\to 2}\left(\frac{\sin(2 - x)}{4 - x^2}\right)$ (iv) $\displaystyle\lim_{n\to\infty}\left(\frac{n + 1}{n^2 + 1}\right)$

   (v) $\displaystyle\lim_{n\to\infty}\frac{(n + 1)(2n + 3)}{(n^2 + 1)}$

1.9 Show that the function $f(x) = x^3 - 1/x$ is continuous at $x = 1$. Discuss the behaviour of the function at $x = 0$.

1.10 Determine the behaviour of $f(x)$ as $x \to \infty$ in the following cases
   (i) $f(x) = e^{-x}(\cosh x + \sin x \sinh x)$

   (ii) $f(x) = \dfrac{(1 + x + x^2)^{1/2}}{x^3}$

26

2
Differential Calculus

2.1 Basic Ideas of Differentiation

The ideas of differential calculus like so many other mathematical discoveries, cannot be attributed to one particular man. They are the results of the work of several mathematicians over a long period. However the subject in its present form is due to two famous seventeenth century mathematicians, Isaac Newton (1642–1727) and Gottfried Willhelm Leibnitz (1646–1716).

Newton, working in England, developed his ideas as a means of describing mechanical systems, while Leibnitz, working in Germany, almost simultaneously and quite independently, used similar ideas to discuss the behaviour of geometric curves.

The basic concept of differentiation, or rate of change, can be seen by considering the geometry of a simple curve. This is the idea introduced by Leibnitz.

Suppose that P and Q are two points on the curve represented by $y = f(x)$ (see Figure 2.1).

Let the x coordinates of P and Q be x_P and x_Q, respectively. The slope of the chord PQ is then

$$\tan \psi_{PQ} = \frac{f(x_Q) - f(x_P)}{x_Q - x_P} \tag{2.1.1}$$

If the point Q is now allowed to approach the point P, the chord PQ, in the limit when Q reaches P, becomes the tangent to the curve at the point P. Hence if ψ is the angle between the tangent to the curve at P and the x-axis

$$\tan \psi = \lim_{Q \to P} \tan \psi_{PQ}$$

$$= \lim_{Q \to P} \frac{f(x_Q) - f(x_P)}{x_Q - x_P} \tag{2.1.2}$$

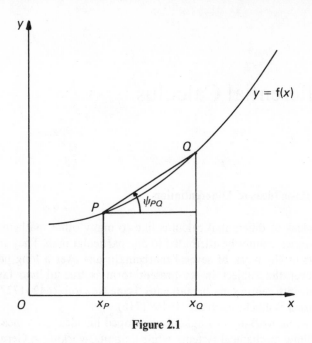

Figure 2.1

Thus the slope or gradient of the curve can be calculated by this limiting process and although $f(x_Q) - f(x_P) \to 0$ and $x_Q - x_P \to 0$ the limit of the quotient is in general well defined.

The slope of the curve, or $\tan \psi$, measures the rate of change of $f(x)$ with respect to x, evaluated at the point P.

There still remains the problem of calculating this quantity for a given functional relationship $y = f(x)$. Formal methods, for a selection of elementary functions, will be given in a later section, but the following example illustrates the process in operation.

Example Given that $f(x) = x^3$, calculate the slope of the curve $y = f(x)$ at the point x.

Let P be the point with coordinate x and let Q have coordinate $x + h$, then

$$\tan \psi = \lim_{Q \to P} \left\{ \frac{f(x + h) - f(x)}{h} \right\}$$

$$= \lim_{h \to 0} \left\{ \frac{(x + h)^3 - x^3}{h} \right\}$$

$$\tan \psi = \lim_{h \to 0} \left\{ \frac{x^3 + 3x^2h + 3xh^2 + h^3 - x^3}{h} \right\}$$

$$= \lim_{h \to 0} \{3x^2 + 3xh + h^2\} = 3x^2$$

This process is called differentiation.

The ideas introduced in this section can now be set out formally without appealing to the geometry of curves.

2.2. Differential Coefficient

DEFINITION   If $f(x)$ is a continuous function of x, then provided that the limit exists, and is independent of the sign of h, the differential coefficient of $f(x)$ at the point $x = a$, is

$$f'(a) = \lim_{h \to 0} \left\{ \frac{f(a + h) - f(a)}{h} \right\} \qquad (2.2.1)$$

DEFINITION   If $f'(x)$ is the function that takes the value $f'(a)$ at each point $x = a$, then $f'(x)$ is called the derivative of $f(x)$ with respect to x and the process of going from $f(x)$ to $f'(x)$ is called differentiation.

A function whose differential coefficient exists at all points is said to be differentiable at all points. If a function is differentiable at a point then it is necessarily continuous at that point. However the converse is not true. There are many functions that are continuous but not differentiable at certain points.

Example   Show that the continuous function $f(x) = |x|$ has no differential coefficient, or is not differentiable at the point $x = 0$.

By formal application of the Definition (2.2.1) for the differential coefficient

$$f'(0) = \lim_{h \to 0} \left\{ \frac{|0 + h| - |0|}{h} \right\}$$

$$= \lim_{h \to 0} \frac{|h|}{h}$$

This limit depends on the sign of h, since, for $h > 0$,

$$\lim_{h \to 0} \frac{|h|}{h} = \lim_{h \to 0} \frac{h}{h} = \lim_{h \to 0} 1 = 1$$

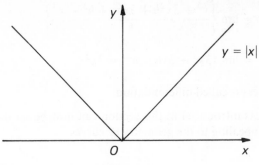

Figure 2.2

and for $h < 0$,

$$\lim_{h \to 0} \frac{|h|}{h} = \lim_{h \to 0} -\frac{h}{h} = \lim_{h \to 0} -1 = -1$$

Hence

$$\lim_{h \to 0^+} \frac{|h|}{h} \neq \lim_{h \to 0^-} \frac{|h|}{h}$$

and therefore $f'(0)$ is not well defined and the function has no differential coefficient at $x = 0$.

The geometrical interpretation of this can be seen from the graph of $y = |x|$ (Figure 2.2).

For $x < 0$, the graph is the straight line $y = -x$, with slope -1, and for $x > 0$ the graph is the line $y = x$, with slope. $+1$. At the point $x = 0$ where these two sections of the graph meet there is no well defined slope.

2.3 Notation

The notation given in Section 2.2 is not the only one used in the differential calculus and it is useful to mention some of the others. Leibnitz originally introduced the following notation.

Suppose that P is the point (x, y) on the curve $y = f(x)$ and that Q is the point $(x + \delta x, y + \delta y)$ on the same curve, then

$$f'(x) = \lim_{\delta x \to 0} \left\{ \frac{f(x + \delta x) - f(x)}{\delta x} \right\} \qquad (2.3.1)$$

$$= \lim_{\delta x \to 0} \left(\frac{\delta y}{\delta x} \right) \qquad (2.3.2)$$

30

which is written as

$$f'(x) = \frac{dy}{dx}.$$ (2.3.3)

In (2.3.3) the quantities dx, dy have no separate meaning. The expression d/dx is an operator which gives the derivative when it is applied to $f(x)$.

It is common practice to abbreviate this notation and write

$$\frac{d}{dx} = D \quad \text{or} \quad \frac{dy}{dx} = Dy$$ (2.3.4)

There are other notations for the derivative (for example y', $\dot{y}$, $Df(x)$, $(d/dx)f(x)$). The two given above are the most widely used but for convenience one or other of the alternatives may be used in later sections. There should be no confusion over meaning.

2.4 Rules for Differentiation

RULE 1   If $f(x) = \lambda F(x) + \mu G(x)$, where λ and μ are constants then

$$f'(x) = \lambda F'(x) + \mu G'(x)$$ (2.4.1)

By formal application of the Definition (2.2.1)

$$f'(x) = \lim_{h \to 0} \left\{ \frac{f(x+h) - f(x)}{h} \right\}$$

$$= \lim_{h \to 0} \left\{ \frac{\lambda F(x+h) + \mu G(x+h) - \lambda F(x) - \mu G(x)}{h} \right\}$$

$$= \lim_{h \to 0} \left\{ \lambda \left[\frac{F(x+h) - F(x)}{h} \right] + \mu \left[\frac{G(x+h) - G(x)}{h} \right] \right\}$$

$$= \lambda \lim_{h \to 0} \left\{ \frac{F(x+h) - F(x)}{h} \right\} + \mu \lim_{h \to 0} \left\{ \frac{G(x+h) - G(x)}{h} \right\}$$

$$= \lambda F'(x) + \mu G'(x)$$

This result means that the derivative of a function which is the sum of several simpler functions can be found by differentiating each element separately and then summing the respective derivatives.

RULE 2   If $f(x) = F(x)G(x)$ then, provided that $F(x)$ and $G(x)$ are differentiable

$$f'(x) = F'(x)G(x) + F(x)G'(x)$$ (2.4.2)

31

Differentiating formally using the Definition (2.2.1)

$$f'(x) = \lim_{h \to 0} \left\{ \frac{F(x + h)G(x + h) - F(x)G(x)}{h} \right\}$$

$$= \lim_{h \to 0} \left\{ \frac{[F(x + h) - F(x)]G(x + h) + F(x)[G(x + h) - G(x)]}{h} \right\}$$

$$= \lim_{h \to 0} \left\{ \left[\frac{F(x + h) - F(x)}{h} \right] G(x + h) \right.$$

$$+ \left. F(x) \left[\frac{G(x + h) - G(x)}{h} \right] \right\}$$

$$= \lim_{h \to 0} \left[\frac{F(x + h) - F(x)}{h} \right] \lim_{h \to 0} G(x + h)$$

$$+ F(x) \lim_{h \to 0} \left[\frac{G(x + h) - G(x)}{h} \right]$$

$$= F'(x)G(x) + F(x)G'(x)$$

This rule for the differentiation of a product requires the two functions $F(x)$ and $G(x)$ to be differentiable. It is possible, however, that a product of two functions is differentiable when the separate functions are not differentiable, although in that situation the product rule does not apply.

For example, consider the function $f(x) = x|x|$, which is the product of two functions $F(x) = x$ and $G(x) = |x|$. Although the function $G(x) = |x|$ is not differentiable at $x = 0$, as was shown by example in section 2.2, the combined function $f(x)$ is differentiable at that point.

This is simply demonstrated by formal application of the definition,

$$f'(0) = \lim_{h \to 0} \left\{ \frac{f(h) - f(0)}{h} \right\}$$

$$= \lim_{h \to 0} \left\{ \frac{h|h| - 0}{h} \right\}$$

$$= \lim_{h \to 0} |h|$$

$$= 0$$

Hence the differential coefficient is defined at $x = 0$ and therefore $f(x)$ is differentiable at $x = 0$ and indeed at all other points.

RULE 3 If $f(x) = F(x)/G(x)$ where $F(x)$ and $G(x)$ are both differentiable and $G(x) \neq 0$ then

$$f'(x) = \frac{G(x)F'(x) - F(x)G'(x)}{G^2(x)} \qquad (2.4.3)$$

This result can be proved by direct application of the definition but this is left as an exercise for the reader. A much simpler proof follows by applying Rule 2 for products.

If $f(x) = F(x)/G(x)$ then $F(x) = f(x)G(x)$.

Differentiation of $F(x)$ using Rule 2 gives

$F'(x) = f'(x)G(x) + f(x)G'(x)$

therefore

$$G(x)f'(x) = F'(x) - f(x)G'(x)$$

$$= F'(x) - \frac{F(x)}{G(x)} G'(x)$$

$$= \frac{G(x)F'(x) - F(x)G'(x)}{G(x)}$$

Dividing by $G(x) \neq 0$

$$f'(x) = \frac{G(x)F'(x) - F(x)G'(x)}{G^2(x)}$$

RULE 4   If z is a function of y and y is in turn a function of a third variable x, then z can be expressed as a function of x and as such, can be differentiated with respect to x. Using the Leibnitz notation this derivative can be written as

$$\frac{dz}{dx} = \frac{dz}{dy} \cdot \frac{dy}{dx} \qquad (2.4.4)$$

Writing $z = z(y)$ and $y = y(x)$

$$\frac{dz}{dx} = \lim_{h \to 0} \left\{ \frac{z\big(y(x+h)\big) - z\big(y(x)\big)}{h} \right\}$$

$$= \lim_{h \to 0} \left\{ \left[\frac{z\big(y(x+h)\big) - z\big(y(x)\big)}{y(x+h) - y(x)} \right] \cdot \left[\frac{y(x+h) - y(x)}{h} \right] \right\}$$

$$= \lim_{h \to 0} \left[\frac{z\big(y(x+h)\big) - z\big(y(x)\big)}{y(x+h) - y(x)} \right] \lim_{h \to 0} \left[\frac{y(x+h) - y(x)}{h} \right] \quad (2.4.5)$$

Let $k = y(x+h) - y(x)$ then as h tends to zero k also tends to zero.

Therefore (2.4.5) can be written as

$$\frac{dz}{dx} = \lim_{k \to 0} \left[\frac{z(y+k) - z(y)}{k} \right] \lim_{h \to 0} \left[\frac{y(x+h) - y(x)}{h} \right]$$

$$= \frac{dz}{dy} \cdot \frac{dy}{dx}$$

A useful consequence of this rule is the derivative of the inverse function. When a functional relationship $y = f(x)$ is one – one, that is each value of x fixes one value of y and vice-versa then x can be expressed as a function of y and can be written as $x = g(y)$. The function g is the inverse of f.

By setting $z = x$ in (2.4.4) it is seen that

$$\frac{dx}{dx} = \frac{dx}{dy} \cdot \frac{dy}{dx}$$

But $dx/dx = 1$, therefore

$$\frac{dx}{dy} \cdot \frac{dy}{dx} = 1$$

and consequently

$$\frac{dx}{dy} = 1 \bigg/ \frac{dy}{dx} \tag{2.4.6}$$

Rule 4 can be expressed in parametric form. If x and y are both functions of a single parameter t then

$$\frac{dy}{dx} = \frac{dy}{dt} \bigg/ \frac{dx}{dt} \tag{2.4.7}$$

2.5 Differentiation of Some Elementary Functions

So far all the rules are general, but in order to apply the differential calculus it is useful to know the derivatives of some simple functions. The results in this section will be obtained formally but will serve as a reference table for later use.

Rational Powers of x

If p is a rational number then

$$\frac{d}{dx} x^p = p x^{p-1} \tag{2.5.1}$$

In the first instance the result is proved for the case $p = n$ when n is a positive integer.

Let $p = n$, then it is necessary to show that

$$\frac{\mathrm{d}x^n}{\mathrm{d}x} = nx^{n-1} \tag{2.5.2}$$

for all integers $n > 0$. For $n = 1$, formal application of the definition gives

$$\frac{\mathrm{d}x^n}{\mathrm{d}x} = \frac{\mathrm{d}x}{\mathrm{d}x} = \lim_{h \to 0} \left\{ \frac{x + h - x}{h} \right\}$$

$$= \lim_{h \to 0} \frac{h}{h}$$

$$= 1$$

That is

$$\frac{\mathrm{d}x^1}{\mathrm{d}x} = 1 \cdot x^{1-1} \tag{2.5.3}$$

and the result (2.5.2) is true for the case $n = 1$.

The Proposition (2.5.2) can now be proved generally by the method of induction. Assume that the result is true for $n = N$ say; that is

$$\frac{\mathrm{d}}{\mathrm{d}x} x^N = Nx^{N-1} \tag{2.5.4}$$

Then it follows that

$$\frac{\mathrm{d}}{\mathrm{d}x} x^{N+1} = \frac{\mathrm{d}}{\mathrm{d}x} x \cdot x^N$$

$$= x \frac{\mathrm{d}}{\mathrm{d}x} x^N + x^N \frac{\mathrm{d}x}{\mathrm{d}x} \quad \text{(product rule)}$$

$$= x \cdot Nx^{N-1} + x^N \cdot 1 \quad \left(\text{using (2.5.3) and (2.5.4)} \right)$$

$$= (N + 1)x^N \tag{2.5.5}$$

Hence if (2.5.2) is true for $n = N$ it is also true for $n = N + 1$. It has already been shown that the result holds for $n = 1$ and therefore must hold for $n = 2$, and similarly for $n = 3, 4, 5$, etc. Thus by the principle of mathematical induction (2.5.2) is true for all values of n.

The proof can now be extended to cover the case of negative integers. In this case $p = -m$ where m is a positive integer.

Let

$$y = x^p = x^{-m} \tag{2.5.6}$$

then

$$x^m y = 1 \tag{2.5.7}$$

Differentiation of both sides of this equation with respect to x using the product rule, gives

$$x^m \frac{dy}{dx} + y \frac{d}{dx} x^m = 0 \tag{2.5.8}$$

Now m is a positive integer so that $d(x^m)/dx = mx^{m-1}$ and (2.5.8) becomes

$$x^m \frac{dy}{dx} + ymx^{m-1} = 0$$

giving

$$\frac{dy}{dx} = -myx^{-1} \tag{2.5.9}$$

Hence, substituting for y from (2.5.6), it follows that

$$\frac{d}{dx} x^{-m} = -mx^{-m-1}$$

and (2.5.1) is seen to be true for $p = -m$.

Thus it follows that for all integer values of p

$$\frac{d}{dx} x^p = px^{p-1} \tag{2.5.10}$$

It is now a simple matter to complete the proof for rational values of p.

If p is rational then it can be expressed as the ratio of two integers, that is in the form $p = n/m$.

Put

$$z = x^p = x^{n/m} \tag{2.5.11}$$

then

$$z^m = x^n \tag{2.5.12}$$

Differentiation of both sides with respect to x using the result already proved for positive integers, and Rule 4 for differentiation gives

$$m \cdot z^{m-1} \frac{dz}{dx} = nx^{n-1}$$

z^m is a function of [handwritten]

Therefore

$$\frac{dz}{dx} = \frac{n}{m} x^{n-1} z^{1-m}$$

$$= \frac{n}{m} x^{n-1} \cdot x^{(1-m)n/m}$$

$$= \frac{n}{m} x^{(n/m)-1}$$

$$= px^{p-1}$$

Hence for all rational values of p

$$\frac{d}{dx} x^p = px^{p-1}$$

The result has now been proved for all rational values of p, but is not in fact limited to these values, indeed (2.5.1) is correct even when p is irrational.

Examples Find the derivatives of the following functions:

(i) $x^4 + 3x + 7$ (ii) $\dfrac{x+1}{x^2+x+1}$ (iii) $x^{1/2}$ (iv) $(1+x^2)^{1/2}$

(i) Let $y = x^4 + 3x + 7$
then

$$\frac{dy}{dx} = \frac{d}{dx}(x^4) + \frac{d}{dx}(3x) + \frac{d}{dx}(7)$$

$$= 4x^3 + 3 + 0$$

that is

$$\frac{d}{dx}(x^4 + 3x + 7) = 4x^3 + 3$$

(ii) Let $y = \dfrac{x+1}{x^2+x+1}$

Applying the quotient Rule 3, for differentiation

$$\frac{dy}{dx} = \frac{(x^2 + x + 1)\dfrac{d}{dx}(x + 1) - (x + 1)\dfrac{d}{dx}(x^2 + x + 1)}{(x^2 + x + 1)^2}$$

$$= \frac{(x^2 + x + 1) \cdot 1 - (x + 1)(2x + 1)}{(x^2 + x + 1)^2}$$

$$= -\frac{x^2 + 2x}{(x^2 + x + 1)^2}$$

(iii) Let $y = x^{1/2}$
then

$$\frac{dy}{dx} = \frac{1}{2} x^{(1/2)-1} = \frac{1}{2} x^{-1/2}$$

Hence

$$\frac{d}{dx} x^{1/2} = \frac{1}{2} x^{-1/2}$$

(iv) Let $z = (1 + x^2)^{1/2}$, then $z = y^{1/2}$ where $y = (1 + x^2)$. Differentiation of z with respect to x using Rule 4 gives

$$\frac{dz}{dx} = \frac{dz}{dy} \cdot \frac{dy}{dx} = \frac{1}{2} y^{-1/2} \frac{dy}{dx} = \tfrac{1}{2}(1 + x^2)^{-1/2} \cdot 2x$$

Therefore

$$\frac{d}{dx}(1 + x^2)^{1/2} = x(1 + x^2)^{-1/2}$$

Trigonometric Functions

The differential coefficients of the two basic functions $\sin x$ and $\cos x$ are obtained by direct application of the definition as follows

$$\frac{d}{dx} \sin x = \lim_{h \to 0} \left\{ \frac{\sin(x + h) - \sin x}{h} \right\}$$

$$= \lim_{h \to 0} \left\{ \frac{1}{h} \cdot 2\cos(x + h/2)\sin h/2 \right\}$$

$$= \lim_{h \to 0} \left\{ \frac{\sin h/2}{h/2} \right\} \lim_{h \to 0} \left\{ \cos(x + h/2) \right\}$$

As $h \to 0$ the first of these two limits is unity and the derivative becomes

$$\frac{\mathrm{d}}{\mathrm{d}x} \sin x = \cos x \qquad (2.5.13)$$

Similarly

$$\frac{\mathrm{d}}{\mathrm{d}x} \cos x = \lim_{h \to 0} \left\{ \frac{\cos(x + h) - \cos x}{h} \right\}$$

$$= \lim_{h \to 0} \left\{ \frac{1}{h} (-2 \sin(x + h/2) \cdot \sin h/2) \right\}$$

$$= -\lim_{h \to 0} \left\{ \frac{\sin h/2}{h/2} \right\} \cdot \lim_{h \to 0} \left\{ \sin(x + h/2) \right\}$$

$$= -1 \cdot \sin x$$

that is

$$\frac{\mathrm{d}}{\mathrm{d}x} \cos x = -\sin x \qquad (2.5.14)$$

The derivatives of the four remaining trigonometric functions are now obtained by using the rules established in Section 2.4.

$$\frac{\mathrm{d}}{\mathrm{d}x} \tan x = \frac{\mathrm{d}}{\mathrm{d}x} \frac{\sin x}{\cos x} = \frac{\cos x \dfrac{\mathrm{d}}{\mathrm{d}x} \sin x - \sin x \dfrac{\mathrm{d}}{\mathrm{d}x} \cos x}{\cos^2 x}$$

$$= \frac{\cos^2 x + \sin^2 x}{\cos^2 x} = \frac{1}{\cos^2 x}$$

that is

$$\frac{\mathrm{d}}{\mathrm{d}x} \tan x = \sec^2 x \qquad (2.5.15)$$

$$\frac{\mathrm{d}}{\mathrm{d}x} \operatorname{cosec} x = \frac{\mathrm{d}}{\mathrm{d}x} (\sin x)^{-1}$$

$$= -(\sin x)^{-2} \frac{\mathrm{d}}{\mathrm{d}x} (\sin x)$$

$$= -\frac{1}{\sin^2 x} \cdot \cos x$$

$$= -\operatorname{cosec} x \cot x \qquad (2.5.16)$$

$$\frac{d}{dx}\sec x = \frac{d}{dx}(\cos x)^{-1}$$

$$= -(\cos x)^{-2}\frac{d}{dx}(\cos x)$$

$$= -\frac{1}{\cos^2 x}\cdot(-\sin x)$$

$$= \sec x \tan x \qquad (2.5.17)$$

$$\frac{d}{dx}\cot x = \frac{d}{dx}(\tan x)^{-1}$$

$$= -(\tan x)^{-2}\frac{d}{dx}(\tan x)$$

$$= -\frac{1}{\tan^2 x}\cdot\sec^2 x$$

$$= -\operatorname{cosec}^2 x \qquad (2.5.18)$$

These results can now be summarized in the following list

$$\frac{d}{dx}\sin x = \cos x$$

$$\frac{d}{dx}\cos x = -\sin x$$

$$\frac{d}{dx}\tan x = \sec^2 x$$

$$\frac{d}{dx}\operatorname{cosec} x = -\operatorname{cosec} x \cot x$$

$$\frac{d}{dx}\sec x = \sec x \tan x$$

$$\frac{d}{dx}\cot x = -\operatorname{cosec}^2 x$$

Inverse Trigonometric Functions

The inverse trigonometric functions as defined in Section 1.5 can now be differentiated by direct application of the results obtained above.

40

Let $y = \sin^{-1} x$, then

$$x = \sin y \qquad (2.5.19)$$

Differentiating both sides of this equation with respect to x

$$1 = \cos y \cdot \frac{dy}{dx}$$

Hence

$$\frac{dy}{dx} = \frac{1}{\cos y}$$

$$= \frac{1}{(1 - \sin^2 y)^{1/2}}$$

$$= \frac{1}{(1 - x^2)^{1/2}}$$

Therefore, substituting back for y

$$\frac{d}{dx} \sin^{-1} x = \frac{1}{(1 - x^2)^{1/2}} \qquad (2.5.20)$$

Similarly if $y = \cos^{-1} x$ then

$$x = \cos y$$

Differentiating with respect to x

$$1 = -\sin y \cdot \frac{dy}{dx}$$

Hence

$$\frac{dy}{dx} = -\frac{1}{\sin y} = -\frac{1}{(1 - \cos^2 y)^{1/2}}$$

giving

$$\frac{d}{dx} \cos^{-1} x = -\frac{1}{(1 - x^2)^{1/2}} \qquad (2.5.21)$$

Finally, if $y = \tan^{-1} x$, then

$$x = \tan y$$

Differentiating with respect to x

$$1 = \sec^2 y \frac{dy}{dx}$$

$$\frac{dx}{dx} = \frac{dy}{dx} \cdot \frac{dx}{dy}$$

$$1 = \frac{d}{dy} \sin y \cdot \frac{dy}{dx}$$

$$= \cos y \frac{dy}{dx}$$

$$\frac{dy}{dx} = \frac{1}{\cos y}$$

41

Hence

$$\frac{dy}{dx} = \frac{1}{\sec^2 y}$$

$$= \frac{1}{1 + \tan^2 y}$$

$$= \frac{1}{1 + x^2}$$

That is

$$\frac{d}{dx} \tan^{-1} x = \frac{1}{1 + x^2} \qquad (2.5.22)$$

Therefore, summarizing the results of this section,

$$\frac{d}{dx} \sin^{-1} x = \frac{1}{(1 - x^2)^{1/2}}$$

$$\frac{d}{dx} \cos^{-1} x = -\frac{1}{(1 - x^2)^{1/2}}$$

$$\frac{d}{dx} \tan^{-1} x = \frac{1}{1 + x^2}$$

Examples Differentiate the following functions with respect to x
(i) $\cos 2x$ (ii) $\cos^3 x$ (iii) $\tan^{-1}(1 + x^2)$

(i) Let $y = \cos 2x$, then $y = \cos u$ where $u = 2x$. Differentiating y with respect to x, using Rule 4 (2.4.4),

$$\frac{dy}{dx} = \frac{dy}{du} \cdot \frac{du}{dx}$$

$$= -\sin u \cdot 2$$

giving

$$\frac{d}{dx} \cos 2x = -2 \sin 2x$$

This result is a special case of the more general result

$$\frac{d}{dx} \cos ax = -a \sin ax$$

where a is a constant.

42

(ii) $y = \cos^3 x$

Let $u = \cos x$ then $y = u^3$. Differentiating with respect to x,

$$\frac{dy}{dx} = 3u^2 \cdot \frac{du}{dx} = 3 \cdot \cos^2 x (-\sin x)$$

Hence

$$\frac{d}{dx} \cos^3 x = -3 \cos^2 x \sin x$$

(iii) $y = \tan^{-1}(1 + x^2)$

Put $u = 1 + x^2$ then $y = \tan^{-1} u$. Differentiating with respect to x,

$$\frac{dy}{dx} = \frac{dy}{du} \cdot \frac{du}{dx}$$

$$= \frac{1}{1 + u^2} \cdot \frac{d}{dx} (1 + x^2)$$

$$= \frac{1}{1 + (1 + x^2)^2} \cdot 2x$$

Hence

$$\frac{d}{dx} \tan^{-1}(1 + x^2) = \frac{2x}{2 + 2x^2 + x^4}$$

With practice the reader will find that many of the steps included in these examples may be omitted.

The Logarithmic Function

The derivative of the logarithmic function follows directly from the formal definition.

Let a be a constant then

$$\frac{d}{dx} (\log_a x) = \lim_{h \to 0} \left\{ \frac{\log_a(x + h) - \log_a x}{h} \right\}$$

$$= \lim_{h \to 0} \frac{1}{h} \log_a \left(\frac{x + h}{x} \right)$$

$$= \lim_{h \to 0} \frac{1}{h} \cdot \log_a \left(1 + \frac{h}{x} \right)$$

$$= \lim_{h \to 0} \frac{1}{x} \cdot \frac{x}{h} \log_a \left(1 + \frac{h}{x} \right)$$

Hence by writing $\alpha = x/h$

$$\frac{d}{dx}(\log_a x) = \frac{1}{x} \cdot \lim_{\alpha \to \infty}\left\{\log_a\left(1 + \frac{1}{\alpha}\right)^\alpha\right\}$$

$$= \frac{1}{x}\log_a\left\{\lim_{\alpha \to \infty}\left(1 + \frac{1}{\alpha}\right)^\alpha\right\} \tag{2.5.23}$$

In Appendix 1.2 (Chapter 1) it was shown that

$$\lim_{\alpha \to \infty}\left(1 + \frac{1}{\alpha}\right)^\alpha = e \tag{2.5.24}$$

where $e = 2.7182818\ldots$, so that

$$\frac{d}{dx}\log_a x = \frac{1}{x}\log_a e \tag{2.5.25}$$

When a takes the value e the resulting logarithmic function, the natural logarithm, is written as

$$\log_e x = \ln x \tag{2.5.26}$$

and its derivative is

$$\frac{d}{dx}\ln x = \frac{1}{x} \cdot \log_e e$$

Therefore, since $\log_e e = 1$,

$$\frac{d}{dx}\ln x = \frac{1}{x} \tag{2.5.27}$$

The Exponential Function

The inverse of the natural logarithm, e^x or $\exp x$ can be differentiated easily by using the result already proved above.

Let $y = e^x$ then

$$x = \ln y \tag{2.5.28}$$

Differentiating with respect to x,

$$1 = \frac{1}{y}\frac{dy}{dx}$$

Therefore

$$\frac{dy}{dx} = y$$

giving

$$\frac{d}{dx} e^x = e^x \qquad (2.5.29)$$

Thus the exponential function has a special property in that the derivative of the function is the function itself.

The Hyperbolic Functions

The hyperbolic functions are simple linear combinations of the exponential function and as such can be easily differentiated as:

$$\frac{d}{dx} \cosh x = \frac{d}{dx} \frac{1}{2} (e^x + e^{-x}) = \frac{1}{2} (e^x - e^{-x}) = \sinh x \qquad (2.5.30)$$

$$\frac{d}{dx} \sinh x = \frac{d}{dx} \frac{1}{2} (e^x - e^{-x}) = \frac{1}{2} (e^x + e^{-x}) = \cosh x \qquad (2.5.31)$$

$$\frac{d}{dx} \tanh x = \frac{d}{dx} \frac{\sinh x}{\cosh x}$$

$$= \frac{\cosh x \cosh x - \sinh x \sinh x}{\cosh^2 x}$$

$$= \operatorname{sech}^2 x \qquad (2.5.32)$$

The Inverse Hyperbolic Functions

The inverse functions play a similar role to the inverse trigonometric functions and can be differentiated in a similar fashion. The proofs of the following results are left as an exercise to the reader.

$$\frac{d}{dx} \sinh^{-1} x = \frac{1}{(x^2 + 1)^{1/2}} \qquad (2.5.33)$$

$$\frac{d}{dx} \cosh^{-1} x = \frac{1}{(x^2 - 1)^{1/2}} \qquad (2.5.34)$$

$$\frac{d}{dx} \tanh^{-1} x = \frac{1}{1 - x^2} \qquad (2.5.35)$$

Examples   Obtain the derivatives of the following functions
(i) $e^{\sin x}$ (ii) $(\ln x)^{3/2}$ (iii) a^x, where a is a constant

(i)   Let $u = \sin x$, then

$$e^{\sin x} = e^u$$

45

and

$$\frac{d}{dx} e^{\sin x} = \frac{d}{dx} e^u$$

$$= e^u \cdot \frac{du}{dx}$$

$$= e^u \cos x$$

that is

$$\frac{d}{dx} e^{\sin x} = \cos x \cdot e^{\sin x}$$

(ii) Put $y = (\ln x)^{3/2}$, then

$$\frac{dy}{dx} = \frac{3}{2}(\ln x)^{1/2} \frac{d}{dx}(\ln x)$$

$$= \frac{3}{2}(\ln x)^{1/2} \cdot \frac{1}{x}$$

Let $\ln x = u$, $y = u^{\frac{3}{2}}$

$$\frac{dy}{dx} = \frac{dy}{du} \cdot \frac{du}{dx}$$

$$= (\frac{d}{du} \cdot u^{\frac{3}{2}})(\frac{d}{dx} \ln x)$$

$$= \frac{3}{2}(u^{\frac{3}{2}-1}) \frac{1}{x}$$

that is

$$\frac{d}{dx}(\ln x)^{3/2} = 3(\ln x)^{1/2}/2x$$

(iii) The function given in this example is simply differentiated by the method known as logarithmic differentiation, a method which has much wider applications.

Let $y = a^x$, then it follows that

$$\ln y = \ln a^x$$

$$= x \ln a$$

Differentiating with respect to x

$$\frac{1}{y}\frac{dy}{dx} = \ln a$$

giving

$$\frac{dy}{dx} = y \ln a$$

Hence

$$\frac{d}{dx}(a^x) = a^x \ln a$$

let $u = \ln y = x \ln a$

$$\frac{du}{dy} \cdot \frac{dy}{dx} = \frac{du}{dx}$$

$$\frac{d}{dy}(\ln y) \cdot \frac{dy}{dx} = \frac{d}{dx}(x \ln a)$$

$$\frac{1}{y} \cdot \frac{dy}{dx} = \ln a$$

$$\frac{dy}{dx} = y \ln a$$

$$\frac{d}{dx} \cdot a^x = a^x \ln a$$

2.6 Higher Derivatives

Since $f'(x)$ is in turn a function of x it too may possess a differential coefficient at the point x. When this derivative exists it is normally written as $f''(x)$. As long as the derivatives exist the process can be repeated and the nth derivative written as $f^{(n)}(x)$, can be found.

The derivatives $f^{(n)}(x)$, $n \geqslant 2$ are known as the higher derivatives. In the Leibnitz notation these are written as

$$\frac{d^2 y}{dx^2}, \frac{d^3 y}{dx^3}, \dots, \frac{d^n y}{dx^n}$$

or

$$D^2 y, D^3 y, \dots, D^n y \tag{2.6.1}$$

2.7 Leibnitz Theorem

The rule for differentiation of a product (2.4.2) can now be extended to include the higher derivatives. The formula was originally obtained by Leibnitz and is expressed as follows

$$\frac{d^n}{dx^n}(F \cdot G) = \frac{d^n F}{dx^n} \cdot G + n \frac{d^{n-1} F}{dx^{n-1}} \cdot \frac{dG}{dx} + \cdots$$

$$+ \frac{n!}{(n-r)! r!} \frac{d^{n-r} F}{dx^{n-r}} \cdot \frac{d^r G}{dx^r} + \cdots + F \cdot \frac{d^n G}{dx^n} \tag{2.7.1}$$

$$= \sum_{r=0}^{n} \frac{n!}{(n-r)! r!} \frac{d^{n-r} F}{dx^{n-r}} \cdot \frac{d^r G}{dx^r} \tag{2.7.2}$$

The proof of this result follows by induction.

From (2.4.2) the theorem is true for $n = 1$. Suppose that the formula holds for $n = N$, then

$$\frac{d^N}{dx^N}(F \cdot G) = \frac{d^N F}{dx^N} \cdot G + N \frac{d^{N-1} F}{dx^{N-1}} \cdot \frac{dG}{dx} + \cdots$$

$$+ \frac{N!}{(N-r+1)!(r-1)!} \frac{d^{N-r+1} F}{dx^{N-r+1}} \cdot \frac{d^{r-1} G}{dx^{r-1}}$$

$$+ \frac{N!}{(N-r)! r!} \frac{d^{N-r} F}{dx^{N-r}} \cdot \frac{d^r G}{dx^r} + \cdots$$

$$+ F \cdot \frac{d^N G}{dx^N} \tag{2.7.3}$$

Differentiating with respect to x, using rule (2.4.2) for products,

$$\frac{d^{N+1}}{dx^{N+1}}(F \cdot G)$$

$$= \frac{d^{N+1}F}{dx^{N+1}} \cdot G + \frac{d^N F}{dx^N} \cdot \frac{dG}{dx} + N\left(\frac{d^N F}{dx^N} \cdot \frac{dG}{dx} + \frac{d^{N-1}F}{dx^{N-1}} \cdot \frac{d^2 G}{dx^2}\right) + \cdots$$

$$+ \frac{N!}{(N-r+1)!(r-1)!}\left(\frac{d^{N-r+2}F}{dx^{N-r+2}} \cdot \frac{d^{r-1}G}{dx^{r-1}} + \frac{d^{N-r+1}F}{dx^{N-r+1}} \cdot \frac{d^r G}{dx^r}\right) +$$

$$+ \frac{N!}{(N-r)!r!}\left(\frac{d^{N-r+1}F}{dx^{N-r+1}} \cdot \frac{d^r G}{dx^r} + \frac{d^{N-r}F}{dx^{N-r}} \cdot \frac{d^{r+1}G}{dx^{r+1}}\right) + \cdots$$

$$+ \frac{dF}{dx} \cdot \frac{d^N G}{dx^N} + F \cdot \frac{d^{N+1}G}{dx^{N+1}}$$

$$= \frac{d^{N+1}F}{dx^{N+1}} \cdot G + (N+1)\frac{d^N F}{dx^N} \cdot \frac{dG}{dx} + \cdots$$

$$+ \left\{\frac{N!}{(N-r+1)!(r-1)!} + \frac{N!}{(N-r)!r!}\right\}\frac{d^{N-r+1}F}{dx^{N-r+1}} \cdot \frac{d^r G}{dx^r} + \cdots$$

$$+ F \cdot \frac{d^{N+1}G}{dx^{N+1}}$$

Therefore

$$\frac{d^{N+1}}{dx^{N+1}}(FG) = \sum_{r=0}^{N+1} \frac{(N+1)!}{(N+1-r)!r!} \frac{d^{N+1-r}F}{dx^{N+1-r}} \cdot \frac{d^r G}{dx^r} \qquad (2.7.4)$$

This is formula (2.7.2) with $n = N + 1$.

Hence if the formula holds for $n = N$ then it holds for $n = N + 1$. Therefore since the formula holds for $n = 1$ it holds for $n = 2$ and so on for all integers n, and the theorem is proved.

2.8 The Mean Value Theorems

Rolle's Theorem

If $f(x)$ is continuous in $[a, b]$ and differentiable in (a, b) and if $f(a) = f(b) = 0$, then there exists a point ξ in (a, b) such that $f'(\xi) = 0$.

Proof If $f(x)$ satisfies the given conditions, then either $f(x) \equiv 0$ for all x in $[a, b]$ in which case there is nothing to prove or there is at least one point inside the interval (a, b) at which the function assumes either its greatest or least value. There is no loss of generality by assuming that the function assumes its greatest value at some point in (a, b).

Let $f(x)$ take its maximum value at the point $x = \xi$ where $a < \xi < b$.

If h is sufficiently small,

$$f(\xi + h) \leqslant f(\xi) \tag{2.8.1}$$

The function $f(x)$ is differentiable at the point $x = \xi$ and

$$f'(\xi) = \lim_{h \to 0^+} \left\{ \frac{f(\xi + h) - f(\xi)}{h} \right\} \leqslant 0 \qquad \text{from (2.8.1)}$$

Also

$$f'(\xi) = \lim_{h \to 0^-} \left\{ \frac{f(\xi + h) - f(\xi)}{h} \right\} \geqslant 0 \qquad \text{from (2.8.1)}$$

Hence, in order that both conditions be satisfied

$$f'(\xi) = 0 \tag{2.8.2}$$

and the theorem is proved.

The Mean Value Theorem

If $f(x)$ is continuous in $[a, b]$ and differentiable in (a, b) then there exists a point ξ in (a, b) such that

$$\frac{f(b) - f(a)}{b - a} = f'(\xi) \tag{2.8.3}$$

Proof   Let $\phi(x) = f(x) - \left\{ f(a) + \dfrac{(x-a)[f(b) - f(a)]}{b-a} \right\}$

$$\phi(x) = f(x) - f(a) - \frac{(x - a)[f(b) - f(a)]}{b - a} \tag{2.8.4}$$

then $\phi(x)$ satisfies the conditions for Rolle's theorem, namely $\phi(x)$ is differentiable in (a, b) and $\phi(a) = \phi(b) = 0$. Therefore there exists a point ξ in (a, b) such that

$$\phi'(\xi) = 0 \qquad \phi(x) = f'(x) - f'(a) \tag{2.8.5}$$

Hence, at the point ξ

$$0 = f'(\xi) - \frac{[f(b) - f(a)]}{b - a}$$

or

$$f'(\xi) = \frac{f(b) - f(a)}{b - a}$$

Generalized Mean Value Theorem

If $f(x)$ and $g(x)$ are two functions, continuous in $[a, b]$ and differentiable in (a, b) then there exists a point ξ in (a, b) such that

$$\frac{f(b) - f(a)}{g(b) - g(a)} = \frac{f'(\xi)}{g'(\xi)} \tag{2.8.6}$$

Proof Let

$$\Psi(x) = [f(b) - f(a)][g(x) - g(a)] \\ - [f(x) - f(a)][g(b) - g(a)]$$

then, clearly $\Psi(x)$ satisfies the conditions for Rolle's theorem. Therefore, there exists a point ξ in (a, b) such that

$$\Psi'(\xi) = 0$$

Hence

$$[f(b) - f(a)]g'(\xi) - f'(\xi)[g(b) - g(a)] = 0$$

giving

$$\frac{f(b) - f(a)}{g(b) - g(a)} = \frac{f'(\xi)}{g'(\xi)}$$

2.9 L'Hopital's Rule

Consider the function $F(x) = f(x)/g(x)$, where $f(x)$ and $g(x)$ are two functions both vanishing when $x = a$. The value of $F(x)$ when $x = a$ is indeterminate, since substitution of that value gives $F(a) = f(a)/g(a) = 0/0$. It may be, however, that $\lim_{x \to a} F(x)$ is well defined.

A typical example of this type of function is $F(x) = \sin x/x$. Here both $\sin x$ and x vanish when $x = 0$, but it was shown geometrically in the previous chapter that $\lim_{x \to 0} F(x) = 1$.

A general rule for evaluating such limits can be deduced as a direct consequence of the generalized mean value theorem.

Given that $f(x)$ and $g(x)$ are two functions that vanish when $x = a$, then for any $x > a$ there exists a value ξ, $a < \xi < x$ such that

$$\frac{f(x) - f(a)}{g(x) - g(a)} = \frac{f'(\xi)}{g'(\xi)} \tag{2.9.1}$$

or

$$\frac{f(x)}{g(x)} = \frac{f'(\xi)}{g'(\xi)} \tag{2.9.2}$$

Now as $x \to a$, $\xi \to a$ therefore when the limit exists

$$\lim_{x \to a} \frac{f(x)}{g(x)} = \lim_{\xi \to a} \frac{f'(\xi)}{g'(\xi)} \tag{2.9.3}$$

This result is known as l'Hopital's rule and is usually written as

$$\lim_{x \to a} \frac{f(x)}{g(x)} = \lim_{x \to a} \frac{f'(x)}{g'(x)} \tag{2.9.4}$$

Example Evaluate

$$\lim_{x \to 0} \left\{ \frac{\cos x - 1}{x^2 - x} \right\}$$

Let $f(x) = \cos x - 1$ and $g(x) = x^2 - x$ then $f(0) = g(0) = 0$ and l'Hopital's rule can be applied, to give

$$\lim_{x \to 0} \left\{ \frac{\cos x - 1}{x^2 - x} \right\} = \lim_{x \to 0} \left\{ \frac{-\sin x}{2x - 1} \right\} = 0$$

L'Hopital's rule can still be applied in cases where $f(x) \to \infty$ and $g(x) \to \infty$ when $x \to a$, simply by writing

$$\lim_{x \to a} \frac{f(x)}{g(x)} = \lim_{x \to a} \frac{1/g(x)}{1/f(x)}$$

Now the functions $1/g(x)$ and $1/f(x)$ both vanish as $x \to a$ and the rule applies. Therefore

$$\begin{aligned}
L = \lim_{x \to a} \frac{f(x)}{g(x)} &= \lim_{x \to a} \frac{1/g(x)}{1/f(x)} \\
&= \lim_{x \to a} \left\{ \frac{-g'(x)}{[g(x)]^2} \middle/ \frac{-f'(x)}{[f(x)]^2} \right\} \\
&= \lim_{x \to a} \left[\frac{f(x)}{g(x)} \right]^2 \left[\frac{g'(x)}{f'(x)} \right] \\
&= \lim_{x \to a} \left(\frac{f(x)}{g(x)} \right)^2 \lim_{x \to a} \left(\frac{g'(x)}{f'(x)} \right) \\
&= L^2 \lim_{x \to a} \frac{g'(x)}{f'(x)}
\end{aligned}$$

Hence

$$\lim_{x \to a} \frac{f(x)}{g(x)} = L = \lim_{x \to a} \frac{f'(x)}{g'(x)} \tag{2.9.5}$$

Similarly, if $f(x)$ and $g(x)$ both tend to zero, or both tend to infinity as x tends to infinity the rule applies. By writing $x = 1/y$.

$$\lim_{x \to \infty} \frac{f(x)}{g(x)} = \lim_{y \to 0} \left\{ \frac{f(1/y)}{g(1/y)} \right\}$$

$$= \lim_{y \to 0} \left\{ -\frac{1}{y^2} f'\left(\frac{1}{y}\right) \middle/ -\frac{1}{y^2} g'(1/y) \right\}$$

$$= \lim_{y \to 0} \left\{ f'(1/y)/g'(1/y) \right\}$$

$$= \lim_{x \to \infty} \frac{f'(x)}{g'(x)}$$

If, after one application of l'Hopital's rule the limit is still indeterminate, the process can be repeated until a determinate form is reached.

Examples Evaluate

(i) $\lim_{x \to 0} \left\{ \dfrac{\sin^2 x}{x^3 + 2x^2} \right\}$ (ii) $\lim_{x \to \infty} x^3 e^{-x^2}$

(i) $\lim_{x \to 0} \left\{ \dfrac{\sin^2 x}{x^3 + 2x^2} \right\} = \lim_{x \to 0} \left\{ \dfrac{2 \sin x \cos x}{3x^2 + 4x} \right\}$

Here, both numerator and denominator still vanish when $x = 0$. Repeating the process

$$\lim_{x \to 0} \left\{ \frac{\sin^2 x}{x^3 + 2x^2} \right\} = \lim_{x \to 0} \left\{ \frac{2 \sin x \cos x}{3x^2 + 4x} \right\}$$

$$= \lim_{x \to 0} \left\{ \frac{2 \cos^2 x - 2 \sin^2 x}{6x + 4} \right\}$$

$$= \frac{2}{4}$$

Therefore

$$\lim_{x \to 0} \left\{ \frac{\sin^2 x}{x^3 + 2x^2} \right\} = \frac{1}{2}$$

(ii) $\lim\limits_{x \to \infty} x^3 e^{-x^2} = \lim\limits_{x \to \infty} \dfrac{x^3}{e^{x^2}}$

$\qquad\qquad = \lim\limits_{x \to \infty} \dfrac{3x^2}{2xe^{x^2}}$

$\qquad\qquad = \lim\limits_{x \to \infty} \dfrac{3x}{2e^{x^2}}$

$\qquad\qquad = \lim\limits_{x \to \infty} \dfrac{3}{4xe^{x^2}} \qquad$ (second application of rule)

$\qquad\qquad = 0$

Exercises

2.1 Differentiate the following functions with respect to x,

(i) $x^3 + x^2 - 5x$ (ii) $3x + \dfrac{1}{x^2} + 1$ (iii) $\dfrac{1}{1 + x^2}$

(iv) $\dfrac{x + 1}{x^2 - x + 1}$ (v) $\dfrac{x}{(x^3 + 1)^{1/2}}$

2.2 Differentiate the following trigonometric functions

(i) $\cos 5x$ (ii) $\dfrac{1 - \cos x}{1 + \cos x}$

(iii) $\sin^2 x \cos^3 x$ (iv) $\dfrac{\cos x - x \sin x}{\sin x + x \cos x}$

(v) $\dfrac{3 + 5 \sin x}{5 + 3 \sin x}$

2.3 Find the derivatives of the functions

(i) $\tan(1/x)$ (ii) $\sin^{-1}\left(\dfrac{3 + 5 \sin x}{5 + 3 \sin x}\right)$

(iii) $[1 + \cos(1 + x)]^3$ (iv) $\sec(1 + x^2)$

(v) $\tan^{-1} x^2$

2.4 Differentiate

(i) $x^2 \sin 2x$ (ii) $\dfrac{\cos x}{1 - x}$ (iii) $\exp(x \cos x)$

with respect to x.

53

2.5 Differentiate the following functions with respect to x

(i) $\sinh(4x + 1)$

(ii) $\cosh(x^2 + 2x)$

(iii) $\tanh x^2$

(iv) $\tanh x \operatorname{sech} 2x$

(v) $\dfrac{\sinh x + \cosh x}{\sinh x - \cosh x}$

2.6 By taking logarithms or otherwise differentiate the functions

$$2^x \qquad 4^{\sqrt{x}} \qquad (\sin x)^{x^2} \qquad (\ln x)^{\ln x} \qquad (x^2 + 1)^x$$

with respect to x.

2.7 If

$$x = \frac{1 + t^2}{1 - t^2} \qquad y = \frac{2t}{1 - t^2}$$

find dy/dx in terms of t and show that

$$\frac{d^2y}{dx^2} = \frac{1}{8}\left(t - \frac{1}{t}\right)^3$$

2.8 If $y = x + e^x$ find d^2x/dy^2 in terms of x.

2.9 Find the slope of the tangent to the curve

$$\frac{x^2}{a^2} + \frac{y^2}{b^2} = 1$$

at the point $(a\cos\theta, b\sin\theta)$.

2.10 Find the nth derivative of x^2e^x.

2.11 By the use of Leibnitz theorem, or otherwise, show that if $f(x)$ satisfies the equation

$$(1 + x^2)f''(x) + xf(x) = 0$$

and the conditions $f(0) = 1$ and $f'(0) = 0$ then $f^{(5)}(0) = 0$ and $f^{(6)}(0) = 28$.

2.12 Evaluate the following limits

(i) $\displaystyle\lim_{x \to 0} \frac{\sin^2 x}{x^2 + x}$

(ii) $\displaystyle\lim_{x \to 0} \frac{\cos x - 1}{\sinh^2 x}$

(iii) $\displaystyle\lim_{x \to 1} \frac{1 - x\sin^2 x}{(1 - x)^2}$

(iv) $\displaystyle\lim_{x \to -1} (1 + x)\ln(1 + x)$

3
Integration

3.1 Indefinite Integral—Antiderivative

The concept of an indefinite integral is very closely linked with the ideas already introduced in the previous chapter. Indeed the process of finding this integral or antiderivative, is the natural inverse operation to differentiation.

DEFINITION If $f(x)$ is a given function of x, and it is possible to construct a function $F(x)$, such that

$$F'(x) = f(x) \tag{3.1.1}$$

then $F(x)$ is called the indefinite integral or antiderivative of $f(x)$ with respect to x.

This indefinite integral is usually written as

$$F(x) = \int f(x)\,dx \tag{3.1.2}$$

In this expression, $f(x)$ is called the integrand.

Clearly if a constant A is added to $F(x)$ then the resulting function

$$G(x) = F(x) + A \tag{3.1.3}$$

also satisfies the condition for the indefinite integral, since

$$G'(x) = F'(x) = f(x) \tag{3.1.4}$$

Thus the operation of finding an indefinite integral, or integration, does not yield a unique result.

Example If $f(x) = x^3$, then using (3.1.1) it can be deduced that

$$F(x) = x^4/4 \tag{3.1.5}$$

satisfies the condition $F'(x) = f(x)$. Hence the indefinite integral of

55

$f(x) = x^3$ is

$$F(x) = x^4/4 + A \tag{3.1.6}$$

where A is any constant..

The constant A is known as the constant of integration.

3.2 Definite Integral—Riemann Integral

The definite integral of a function $f(x)$ in an interval $[a, b]$ can in many cases be defined simply as the area contained by the curve $y = f(x)$, the x axis and the ordinates $x = a$, $x = b$, that is the shaded area shown in Figure 3.1. In some situations however, the geometric interpretation is not meaningful and a formal definition of the integral is essential.

DEFINITION Suppose that $f(x)$ is a function of x defined in the interval $[a, b]$. Let the interval $[a, b]$ be subdivided into N sections by points $P_n(n = 0, 1, 2, 3, \ldots, N)$, with $x = \xi_n$ at the point P_n, such that

$$\xi_0 = a, \, \xi_n < \xi_{n+1} \qquad \text{and} \qquad \xi_N = b \tag{3.2.1}$$

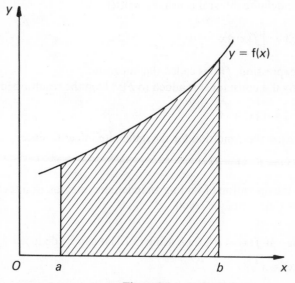

Figure 3.1

Let the sum S_N be defined as

$$S_N = \sum_{n=0}^{N-1} f(\xi_n)(\xi_{n+1} - \xi_n) \tag{3.2.2}$$

$$= \sum_{n=0}^{N-1} f(\xi_n)\Delta\xi_n \tag{3.2.3}$$

where $\Delta\xi_n = \xi_{n+1} - \xi_n$. Then, provided that the limit exists, the definite integral or Riemann integral of $f(x)$ with respect to x, between the points $x = a$ and $x = b$ is defined as $\lim_{N\to\infty} S_N$, and is written as

$$\int_a^b f(x)\,dx = \lim_{N\to\infty} S_N \tag{3.2.4}$$

Alternatively

$$\int_a^b f(x)\,dx = \lim_{\Delta\xi_n\to 0} \sum_{n=0}^{N-1} f(\xi_n)\Delta\xi_n \tag{3.2.5}$$

The interval $[a, b]$ is called the range of integration.

It can now be shown that where both ideas are applicable the formal definition and the simple geometric ideas are equivalent. From Figure 3.2 it can be seen that, in the case where the curve $y = f(x)$ lies

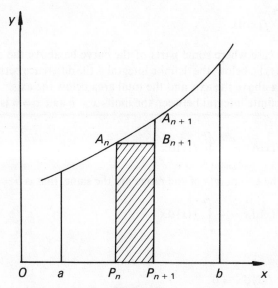

Figure 3.2

above the x axis, that is when

$$f(x) > 0 \qquad \text{for } a < x < b \tag{3.2.6}$$

then

$$f(\xi_n)(\xi_{n+1} - \xi_n) \tag{3.2.7}$$

is the area of the rectangle $A_n P_n P_{n+1} B_{n+1}$ and therefore S_N is the total area contained by all such rectangles. Hence as $N \to \infty$ or $\xi_{n+1} - \xi_n \to 0$, and the points B_{n+1} approach the points A_n, the limit of S_N becomes the area between the curve and the x axis.

Hence

$$\int_a^b f(x)\,dx = \text{area bounded by } y = f(x),\ y = 0$$

$$\text{and the ordinates } x = a,\ x = b \tag{3.2.8}$$

If the curve $y = f(x)$ lies below the x axis, then

$$f(x) < 0 \qquad \text{for } a < x < b \tag{3.2.9}$$

then the definite integral

$$\int_a^b f(x)\,dx$$

is necessarily negative. In this case the area contained by the curve and the x axis is

$$-\int_a^b f(x)\,dx \tag{3.2.10}$$

In the case where some parts of the curve lie above the axis and some parts lie below, the definite integral is the difference between the total area above the axis and the total area below the axis.

The definite integral between the limits $x = b$ and $x = a$ is defined as

$$\int_b^a f(x)\,dx = -\int_a^b f(x)\,dx \tag{3.2.11}$$

Thus if the end points of the range are the same, that is $b = a$

$$\int_a^a f(x)\,dx = -\int_a^a f(x)\,dx \tag{3.2.12}$$

giving

$$2\int_a^a f(x)\,dx = 0$$

Hence

$$\int_a^a f(x)\,dx = 0 \qquad (3.2.12)$$

3.3 Properties of Definite Integrals

(a) $\qquad \displaystyle\int_a^c f(x)\,dx = \int_a^b f(x)\,dx + \int_b^c f(x)\,dx \qquad (3.3.1)$

If $a < b < c$ the result follows immediately from the formal definition of the definite integral.

If $a < c < b$, then the formal definition gives

$$\int_a^b f(x)\,dx = \int_a^c f(x)\,dx + \int_c^b f(x)\,dx$$

that is

$$\int_a^c f(x)\,dx = \int_a^b f(x)\,dx - \int_c^b f(x)\,dx \qquad (3.3.2)$$

Now

$$\int_c^b f(x)\,dx = -\int_b^c f(x)\,dx \qquad (3.3.3)$$

therefore (3.3.2) gives

$$\int_a^c f(x)\,dx = \int_a^b f(x)\,dx + \int_b^c f(x)\,dx \qquad (3.3.4)$$

Hence (3.3.1) is true always.

(b) If $f(x)$ and $g(x)$ are two functions defined in $[a, b]$ and λ, μ are constants

$$\int_a^b \{\lambda f(x) + \mu g(x)\}\,dx = \lambda \int_a^b f(x)\,dx + \mu \int_a^b g(x)\,dx \qquad (3.3.5)$$

This result is a direct consequence of the formal definition.

(c) If m and M are two numbers such that $m \leqslant f(x) \leqslant M$ for all x in $[a, b]$, then

$$m(b - a) \leqslant \int_a^b f(x)\,dx \leqslant M(b - a) \qquad (3.3.6)$$

PROOF   From the definition

$$\int_a^b f(x)\,\mathrm{d}x = \lim_{N\to\infty} \sum_{n=0}^{N-1} f(\xi_n)(\xi_{n+1} - \xi_n)$$

$$\geqslant \lim_{N\to\infty} \sum_{n=0}^{N-1} m(\xi_{n+1} - \xi_n)$$

$$= m \lim_{N\to\infty} \sum_{n=0}^{N-1} (\xi_{n+1} - \xi_n)$$

$$= m \lim_{N\to\infty} (\xi_N - \xi_0)$$

$$= m(b - a)$$

Hence

$$m(b - a) \leqslant \int_a^b f(x)\,\mathrm{d}x$$

By a similar argument it is easily shown that

$$M(b - a) \geqslant \int_a^b f(x)\,\mathrm{d}x$$

Hence

$$m(b - a) \leqslant \int_a^b f(x)\,\mathrm{d}x \leqslant M(b - a)$$

3.4   The Mean Value Theorem of the Integral Calculus

If $f(x)$ is a continuous function of x in the interval $[a, b]$ then there is some value ξ, $a < \xi < b$, for which

$$\int_a^b f(x)\,\mathrm{d}x = (b - a)f(\xi) \tag{3.4.1}$$

PROOF   The function $f(x)$ is continuous in $[a, b]$, therefore it must be bounded in this interval. Let M be the largest value of $f(x)$ and m be the smallest value of $f(x)$ in the interval $[a, b]$.

Then using (3.3.6)

$$m(b - a) \leqslant \int_a^b f(x)\,\mathrm{d}x \leqslant M(b - a)$$

60

that is

$$m \leqslant \frac{\int_a^b f(x)\,dx}{b-a} \leqslant M \qquad (3.4.2)$$

Since $f(x)$ is a continuous function and M, m are its largest and smallest values, respectively, then $f(x)$ must assume every value between M and m for some value of x. Now, from (3.4.2)

$$\frac{\int_a^b f(x)\,dx}{b-a}$$

is a number lying between M and m, therefore for some value $x = \xi$, $a < \xi < b$

$$\frac{\int_a^b f(x)\,dx}{b-a} = f(\xi) \qquad (3.4.3)$$

Therefore

$$\int_a^b f(x)\,dx = (b-a)f(\xi) \qquad (3.4.4)$$

This result may be written alternatively as

$$\int_a^b f(x)\,dx = (b-a)f(a + \theta[b-a]) \qquad 0 < \theta < 1 \qquad (3.4.5)$$

This theorem can be generalized in the following form.
If $g(x)$ is a non-negative continuous function, then since

$$mg(x) \leqslant f(x)g(x) \leqslant Mg(x) \qquad (3.4.6)$$

it follows that

$$m \int_a^b g(x)\,dx \leqslant \int_a^b f(x)g(x)\,dx \leqslant M \int_a^b g(x)\,dx \qquad (3.4.7)$$

Hence, there exists a number ξ, $a < \xi < b$ such that

$$\int_a^b f(x)g(x)\,dx = f(\xi) \int_a^b g(x)\,dx \qquad (3.4.8)$$

61

3.5   Equivalence of the Definite and Indefinite Integral

In Sections 3.1 and 3.2 the term integral has been introduced in two different ways. It would be clearly undesirable for the same name to be given to two completely unrelated objects, and in this section it is shown that in the case of continuous functions when *both* the definite and the indefinite integral are well defined, both definitions are equivalent.

Consider the Riemann integral of a function $f(x)$ over the interval $[c, X]$ where c is fixed, and write

$$F(X) = \int_c^X f(x)\,dx \tag{3.5.1}$$

The value $F(X)$ depends on the end point X, in other words $F(X)$ is a function of X. If $f(x)$ is continuous then this expression can be differentiated formally with respect to X using the Definition (2.2.1) as follows

$$F'(X) = \lim_{h \to 0} \frac{F(X + h) - F(X)}{h}$$

$$= \lim_{h \to 0} \frac{1}{h}\left\{\int_c^{X+h} f(x)\,dx - \int_c^X f(x)\,dx\right\}$$

$$= \lim_{h \to 0} \frac{1}{h}\int_X^{X+h} f(x)\,dx$$

$$= \lim_{h \to 0} \frac{1}{h}\cdot h f(X + \theta h) \qquad \text{where } 0 < \theta < 1$$

$$= \lim_{h \to 0} f(X + \theta h)$$

$$= f(X) \tag{3.5.2}$$

Thus the function $F(X)$ satisfies the condition for the indefinite integral, namely

$$F'(X) = f(X) \tag{3.5.3}$$

and therefore both ideas are equivalent.

This equivalence gives a simple method for calculating definite integrals when $f(x)$ is continuous. Using (3.3.1) the definite integral of

62

the function $f(x)$ over the range $[a, b]$ can be written

$$\int_a^b f(x)\,dx = \int_a^c f(x)\,dx + \int_c^b f(x)\,dx$$

$$= \int_c^b f(x)\,dx - \int_c^a f(x)\,dx$$

where c is any constant.

Hence using the notation from (3.5.1)

$$\int_a^b f(x)\,dx = F(b) - F(a) \tag{3.5.4}$$

It must be remembered that the indefinite integral is not unique, as demonstrated in Section 3.1.

Suppose that $F_1(x)$ and $F_2(x)$ are two indefinite integrals of $f(x)$, then $F_1(x)$ and $F_2(x)$ differ only by a constant, that is

$$F_1(x) - F_2(x) = A \tag{3.5.5}$$

Therefore

$$F_1(b) - F_1(a) = F_2(b) + A - [F_2(a) + A]$$
$$= F_2(b) - F_2(a) \tag{3.5.6}$$

Hence the validity of (3.5.4) does not depend on the choice of a particular integral.

3.6 Integration of Some Simple Functions

Using the results given in Chapter 2 for the derivatives of the elementary functions it is possible to construct a short table of the corresponding integral formulae.

If p is replaced by $p + 1$, the expression (2.5.1) becomes

$$\frac{d}{dx} x^{p+1} = (p + 1)x^p$$

therefore using the idea of the antiderivative it follows that, except when $p = -1$

$$\int x^p\,dx = \frac{1}{p + 1} x^{p+1} + A \tag{3.6.1}$$

where A is the constant of integration.

63

The case $p = -1$, will be considered in Section 3.7.

This process can be extended to include many elementary functions. Some typical results which follow directly as antiderivatives from the corresponding formula for differentiation are

$$\int e^x \, dx = e^x \qquad (3.6.2)$$

$$\int \cos x \, dx = \sin x \qquad (3.6.3)$$

$$\int \sin x \, dx = -\cos x \qquad (3.6.4)$$

$$\int \sec^2 x \, dx = \tan x \qquad (3.6.5)$$

$$\int \operatorname{cosec}^2 x \, dx = -\cot x \qquad (3.6.6)$$

$$\int \cosh x \, dx = \sinh x \qquad (3.6.7)$$

$$\int \sinh x \, dx = \cosh x \qquad (3.6.8)$$

In these formulae the constant of integration has been omitted for simplicity. This will also be the case in the following sections but it should be remembered that the constant is an important feature of the indefinite integral.

3.7 Some Special Methods of Integration

In practice, it is unlikely that an integral is immediately recognizable as an antiderivative. It is therefore necessary to devise techniques by which the integral can be transformed into a form that lends itself to this method of integration. Some of the standard techniques are included in this section.

Method of Substitution

Let $f(x)$ be a given function of x, then from the definition of the integral

$$\int f(x) \, dx = F(x) \qquad (3.7.1)$$

64

where $F(x)$ satisfies the condition

$$F'(x) = f(x) \qquad (3.7.2)$$

If $F(x)$ is not immediately recognizable as an antiderivative it may be possible to transform the integral into a recognizable form by a simple change of variable.

Suppose that x is written in terms of a new variable u by

$$x = x(u) \qquad (3.7.3)$$

then the integrand f and the integral F will be replaced by functions $g(u)$ and $G(u)$ where

$$g(u) = f[x(u)]$$
$$G(u) = F[x(u)] \qquad (3.7.4)$$

With this substitution, using rule (2.4.4) for differentiation (3.7.2) becomes

$$g(u) = \frac{dG}{dx}$$

$$= \frac{dG}{du} \cdot \frac{du}{dx} \qquad (3.7.5)$$

giving

$$\frac{dG}{du} = g(u)\frac{dx}{du} \qquad (3.7.6)$$

Therefore by comparison with (3.7.2), $G(u)$ is the indefinite integral of the function

$$g(u)\frac{dx}{du}$$

and can be written as

$$G(u) = \int g(u)\frac{dx}{du} \cdot du \qquad (3.7.7)$$

This integral, depending on the function $g(u)$, can in many cases be reduced to a recognizable antiderivative by an appropriate choice of the substitution $x = x(u)$. Once the integral has been evaluated and $G(u)$ is known it is a relatively simple matter to substitute for u in terms of x and recover the function $F(x)$.

65

Examples

(i) If a is a constant, determine

$$\int \frac{dx}{a^2 + x^2}$$

Let $x = a \tan u$, then using (3.7.7)

$$\int \frac{dx}{a^2 + x^2} = \int \frac{1}{a^2(1 + \tan^2 u)} \cdot \frac{d}{du}(a \tan u) \, du$$

$$= \int \frac{1}{a^2 \sec^2 u} a \sec^2 u \, du$$

$$= \int \frac{1}{a} \cdot 1 \, du$$

$$= \frac{1}{a} \cdot u$$

Therefore, recovering x as the variable

$$\int \frac{dx}{x^2 + a^2} = \frac{1}{a} \tan^{-1}\left(\frac{x}{a}\right) \qquad (3.7.8)$$

(ii) Find

$$\int \frac{1}{(a^2 - x^2)^{1/2}} \, dx$$

when a is constant. Put $x = a \sin u$, then

$$\int \frac{1}{(a^2 - x^2)^{1/2}} \, dx = \int \frac{1}{a(1 - \sin^2 u)^{1/2}} \frac{d}{du}(a \sin u) \, du$$

$$= \int \frac{1}{a \cos u} \cdot a \cos u \, du$$

$$= \int 1 \cdot du$$

$$= u$$

Hence

$$\int \frac{dx}{(a^2 - x^2)^{1/2}} = \sin^{-1}(x/a) \qquad (3.7.9)$$

This result might have been obtained directly from (2.5.20).

(iii) Find

$$\int \frac{dx}{x}$$

This is the special case $p = -1$ deferred from (3.6.1). When $x > 0$, the integral may be obtained as an antiderivative using (2.5.27), as

$$\int \frac{1}{x} dx = \ln x \qquad (3.7.10)$$

For $x < 0$, put $x = -u$ where $u > 0$, then

$$\int \frac{1}{x} dx = \int -\frac{1}{u} \cdot \frac{d}{du}(-u) \, du$$

$$= \int \frac{1}{u} \cdot du$$

$$= \ln u \qquad \text{since } u > 0$$

$$= \ln(-x) \qquad (3.7.11)$$

These two cases can be combined to give

$$\int \frac{1}{x} dx = \ln |x| \qquad (3.7.12)$$

More generally it can be shown that if $g(x)$ is a differentiable function, then

$$\int \frac{g'(x)}{g(x)} \, dx = \ln |g(x)| \qquad (3.7.13)$$

Two useful results which can be obtained by a simple change of variable are given below.

Suppose that

$$\int f(x) \, dx = F(x)$$

then, if a is a constant,

$$\int f(ax) \, dx = \frac{1}{a} F(ax) \qquad (3.7.14)$$

and

$$\int f(a + x) \, dx = F(a + x) \qquad (3.7.15)$$

67

Integration by Parts

Let $U(x)$ and $V(x)$ be two functions of x, then the product rule for differentiation gives

$$\frac{d}{dx}(UV) = U\frac{dV}{dx} + V\frac{dU}{dx} \qquad (3.7.16)$$

This formula can be rearranged to give

$$U\frac{dV}{dx} = \frac{d}{dx}(UV) - V\frac{dU}{dx} \qquad (3.7.17)$$

This form of the product rule can often be used to advantage in the evaluation of the integral

$$\int f(x)\,dx$$

when the function $f(x)$ is not immediately integrable, but can be written in the form

$$f(x) = U(x)\frac{dV}{dx} \qquad (3.7.18)$$

In this case

$$\int f(x)\,dx = \int U\frac{dV}{dx}\,dx$$

$$= \int \left\{\frac{d}{dx}(UV) - V\frac{dU}{dx}\right\}\,dx$$

$$= UV - \int V\frac{dU}{dx}\,dx \qquad (3.7.19)$$

Examples

(i) $\displaystyle\int x \sin x \, dx$

This integral can be put into a form suitable for integration by parts, by writing $U(x) = x$ and $V(x) = -\cos x$, so that

$$\int x \sin x \, dx = \int x \frac{d}{dx}(-\cos x)\,dx$$

$$= -x\cos x - \int(-\cos x)\cdot 1 \, dx \quad (\text{using } (3.7.19))$$

68

$$= -x \cos x + \int \cos x \, dx$$

$$= -x \cos x + \sin x \qquad (3.7.20)$$

(ii) $\int \ln x \, dx$

By suitable choice of $U(x)$ and $V(x)$, this integral can be integrated by parts as

$$\int \ln x \, dx = \int 1 \cdot \ln x \, dx$$

$$= \int \ln x \cdot \frac{d}{dx} x \, dx$$

$$= x \ln x - \int x \frac{d}{dx} \ln x \, dx$$

$$= x \ln x - \int x \cdot \frac{1}{x} \, dx$$

$$= x \ln x - x$$

Hence

$$\int \ln x \, dx = x(\ln x - 1) \qquad (3.7.21)$$

(iii) $\int e^x \sin x \, dx$

For integration by parts, let $U(x) = \sin x$ and $V(x) = e^x$ then

$$\int e^x \sin x \, dx = \int \sin x \cdot \frac{d}{dx} e^x \, dx$$

$$= \sin x \cdot e^x - \int e^x \frac{d}{dx} (\sin x) \, dx$$

$$= e^x \sin x - \int e^x \cos x \, dx$$

$$= e^x \sin x - \int \frac{d}{dx} (e^x) \cdot \cos x \, dx$$

$$\int e^x \sin x \, dx = e^x \sin x - \left\{ e^x \cos x - \int e^x \frac{d}{dx}(\cos x) \, dx \right\}$$

$$= e^x \sin x - \left\{ e^x \cos x + \int e^x \sin x \, dx \right\}$$

$$= e^x \sin x - e^x \cos x - \int e^x \sin x \, dx \qquad (3.7.22)$$

On inspection of the Equation (3.7.22) it is seen that the original integral has reappeared on the right-hand side. The equation can now be rearranged to give

$$2 \int e^x \sin x \, dx = e^x \sin x - e^x \cos x$$

Hence

$$\int e^x \sin x \, dx = \tfrac{1}{2} e^x (\sin x - \cos x) \qquad (3.7.23)$$

This example is typical of many where repeated integration by parts throws up the original integral, thus allowing that integral to be evaluated by regrouping the equation. More examples of this technique will appear in later sections.

Partial Fractions

This method is applicable when the integrand $f(x)$ can be written as the ratio of two polynomials, so that

$$\int f(x) \, dx = \int \frac{P(x)}{Q(x)} \, dx \qquad (3.7.24)$$

where $P(x)$ and $Q(x)$ are polynomials.

The method requires that the degree of $P(x)$ be less than the degree of $Q(x)$. If this is not the case then the integrand must be divided out and the integral expressed in the form

$$\int f(x) \, dx = \int \left\{ p(x) + \frac{R(x)}{Q(x)} \right\} dx$$

$$= \int p(x) \, dx + \int \frac{R(x)}{Q(x)} \, dx \qquad (3.7.25)$$

where $p(x)$ and $R(x)$ are polynomials, the degree of $R(x)$ being less than that of $Q(x)$.

Examples Express the functions

(i) $x^3/(x^2 + 1)$

(ii) $\dfrac{x^4 + x}{x^2 - 3x + 2}$

in the form $p(x) + R(x)/Q(x)$, where the degree of $R(x)$ is less than that of $Q(x)$.

(i) Here $f(x) = \dfrac{x^3}{x^2 + 1}$

$$= \frac{x(x^2 + 1) - x}{x^2 + 1}$$

$$= x - \frac{x}{x^2 + 1}$$

(ii) $f(x) = \dfrac{x^4 + x}{x^2 - 3x + 2}$

$$= \frac{x^2(x^2 - 3x + 2) + 3x^3 - 2x^2 + x}{x^2 - 3x + 2}$$

$$= x^2 + \frac{3x^3 - 2x^2 + x}{x^2 - 3x + 2}$$

$$= x^2 + \frac{3x(x^2 - 3x + 2) + 9x^2 - 6x - 2x^2 + x}{(x^2 - 3x + 2)}$$

$$= x^2 + 3x + \frac{7x^2 - 5x}{x^2 - 3x + 2}$$

$$= x^2 + 3x + \frac{7(x^2 - 3x + 2) + 21x - 14 - 5x}{x^2 - 3x + 2}$$

$$= x^2 + 3x + 7 + \frac{16x - 14}{x^2 - 3x + 2}$$

$A = \$6 - 13$

$4x \quad A + B = 16$

$\dfrac{14}{A} \quad \dfrac{2}{B} \quad -A - 2B = -14$

$(x-2) \quad (x-1) \quad -16 + B - 213 = 14$

$(x-2)(x-1) \quad A(x-1) + B(x-2) = 16x - 14 \quad 2$

The integration of the polynomial $p(x)$ is straightforward using (3.6.1), therefore there is no loss of generality by considering only integrals of the form $\int P(x)/Q(x)\,dx$ where the degree of $P(x)$ *is* less than that of $Q(x)$.

The following examples serve to illustrate how the method works in some simple cases.

Examples

(i) Express

$$\frac{2x + 1}{(x + 2)(x^2 + 2x - 3)}$$

in terms of partial fractions and then find

$$\int \frac{2x + 1}{(x + 2)(x^2 + 2x - 3)} \, dx$$

The integrand can be written in the form

$$\frac{2x + 1}{(x + 2)(x^2 + 2x - 3)} = \frac{2x + 1}{(x + 2)(x - 1)(x + 3)} \qquad (3.7.26)$$

and hence in partial fractions as

$$\frac{2x + 1}{(x + 2)(x - 1)(x + 3)} \equiv \frac{A}{x + 2} + \frac{B}{x - 1} + \frac{C}{x + 3} \qquad (3.7.27)$$

where A, B and C are constants.

To determine A, B and C multiply both sides of (3.7.27) by $(x + 2)(x - 1)(x + 3)$, to obtain

$$2x + 1 \equiv A(x - 1)(x + 3) + B(x + 2)(x + 3)$$
$$+ C(x + 2)(x - 1) \qquad (3.7.28)$$

This identity must be true for all values of x, therefore the coefficients of the various powers of x on either side of the equation must be identical.

Hence, equating coefficients

$$x^2: \qquad A + B + C = 0$$
$$x : \qquad 2A + 5B + C = 2$$
$$x^0: \qquad -3A + 6B - 2C = 1$$

These equations are easily solved to give

$$A = 1 \qquad B = \tfrac{1}{4} \qquad C = -\tfrac{5}{4}$$

so that

$$\frac{2x + 1}{(x + 2)(x^2 + 2x - 3)} = \frac{1}{4} \left\{ \frac{1}{x - 1} + \frac{4}{x + 2} - \frac{5}{x + 3} \right\}$$

$$(3.7.29)$$

72

The integral can now be written as

$$\int \frac{2x+1}{(x+2)(x^2+2x-3)}\,dx = \frac{1}{4}\int\left\{\frac{1}{x-1}+\frac{4}{x+2}-\frac{5}{x+3}\right\}dx$$

$$= \frac{1}{4}\{\ln|x-1|+4\ln|x+2|$$

$$- 5\ln|x+3|\}$$

$$= \frac{1}{4}\{\ln|x-1|+\ln|x+2|^4$$

$$- \ln|x+3|^5\}$$

$$= \frac{1}{4}\ln\left|\frac{(x+2)^4(x-1)}{(x+3)^5}\right| \quad (3.7.30)$$

(ii) Express $1/x^2(x^2+1)$ in partial fractions and evaluate

$$\int \frac{dx}{x^2(x^2+1)}$$

In this case the function is expressed in partial fractions as

$$\frac{1}{x^2(x^2+1)} \equiv \frac{A}{x}+\frac{B}{x^2}+\frac{Cx+D}{x^2+1} \quad (3.7.31)$$

where A, B, C and D are constants.
Multiply both sides of (3.7.31) by $x^2(x^2+1)$ then

$$1 \equiv Ax(x^2+1) + B(x^2+1) + (Cx+D)x^2$$

Hence, equating the coefficient of the powers of x

$$
\begin{array}{llll}
x^3: & A & +C & =0\\
x^2: & B & +D & =0\\
x: & A & & =0\\
x^0: & B & & =1
\end{array}
$$

so that

$$A=0 \quad B=1 \quad C=0 \quad D=-1$$

Therefore

$$\frac{1}{x^2(x^2+1)} = \frac{1}{x^2}-\frac{1}{x^2+1} \quad (3.7.32)$$

73

and the integral becomes

$$\int \frac{dx}{x^2(x^2+1)} = \int \left\{ \frac{1}{x^2} - \frac{1}{x^2+1} \right\} dx$$

$$= -\frac{1}{x} - \tan^{-1} x \qquad (3.7.33)$$

(iii) Express

$$\frac{x-2}{(x+2)(x^2+2x+2)}$$

in partial fractions and evaluate

$$\int \frac{(x-2)\,dx}{(x+2)(x^2+2x+2)}$$

In terms of partial fractions

$$\frac{x-2}{(x+2)(x^2+2x+2)} \equiv \frac{A}{x+2} + \frac{Bx+C}{x^2+2x+2}$$

where A, B and C are constants.
   Multiplying by $(x+2)(x^2+2x+2)$

$$A(x^2+2x+2) + (Bx+C)(x+2) \equiv x - 2$$

Equating coefficients

$$
\begin{aligned}
x^2: & \quad A + B && = 0 \\
x: & \quad 2A + 2B + C && = 1 \\
x^0: & \quad 2A \quad\quad + 2C && = -2
\end{aligned}
$$

Hence

$$A = -2 \qquad B = 2 \qquad C = 1$$

Therefore

$$\frac{(x-2)}{(x+2)(x^2+2x+2)} = \frac{2x+1}{x^2+2x+2} - \frac{2}{x+2}$$

so that

$$\int \frac{(x-2)\,dx}{(x+2)(x^2+2x+2)}$$

$$= \int \frac{2x+1}{x^2+2x+2}\,dx - 2\int \frac{dx}{x+2}$$

$$= \int \left\{ \frac{2x+2}{x^2+2x+2} - \frac{1}{x^2+2x+2} \right\} dx - 2 \int \frac{dx}{x+2}$$

$$= \ln|x^2+2x+2| - \int \frac{dx}{(x+1)^2+1} - 2\ln|x+2|$$

$$= \ln|x^2+2x+2| - \tan^{-1}(x+1) - \ln|x+2|^2$$

$$= \ln\left| \frac{x^2+2x+2}{(x+2)^2} \right| - \tan^{-1}(x+1)$$

(iv) Evaluate

$$\int \frac{dx}{x^2-a^2},$$

where a is a constant.
The integrand can be expressed in partial fractions as

$$\frac{1}{x^2-a^2} = \frac{1}{2a} \left\{ \frac{1}{x-a} - \frac{1}{x+a} \right\}$$

Therefore

$$\int \frac{1}{x^2-a^2} dx = \frac{1}{2a} \int \left\{ \frac{1}{x-a} - \frac{1}{x+a} \right\} dx$$

$$= \frac{1}{2a} \{\ln|x-a| - \ln|x+a|\}$$

Hence

$$\int \frac{1}{x^2-a^2} dx = \frac{1}{2a} \ln\left| \frac{x-a}{x+a} \right| \qquad (3.7.34)$$

In general when the quotient $R(x)/Q(x)$ is expressed in terms of partial fractions it will appear as a linear combination of terms of one of the following forms;

$$\frac{1}{(x+a)^n} \quad \text{or} \quad \frac{x+d}{x^2+2bx+c} \qquad (3.7.35)$$

where n is a positive integer and a, b, c and d are real constants.
Hence the integrals involved are either,

$$\int \frac{dx}{(x+a)^n} \qquad (3.7.36)$$

or

$$\int \frac{x + d}{x^2 + 2bx + c} \, dx \qquad\qquad (3.7.37)$$

The integral (3.7.36) is simply evaluated by writing $u = x + a$. Then

$$\int \frac{dx}{(x + a)^n} = \int \frac{1}{u^n} \, du$$

$$= -\frac{1}{n - 1} u^{-n+1} \qquad \text{if } n \neq 1$$

$$= \ln |u| \qquad\qquad \text{if } n = 1$$

Hence

$$\int \frac{dx}{(x + a)^n} = \frac{-1}{(n - 1)} (x + a)^{-n+1} \qquad \text{if } n \neq 1 \qquad (3.7.38a)$$

$$= \ln |x + a| \qquad\qquad \text{if } n = 1 \qquad (3.7.38b)$$

The second integral (3.7.37) is more complicated, and some rearrangement is necessary to reduce it to an elementary form. Let

$$I = \int \frac{x + d}{x^2 + 2bx + c} \, dx$$

$$= \int \frac{x + d}{(x + b)^2 + c - b^2} \, dx$$

$$= \int \frac{x + b}{(x + b)^2 + c - b^2} \, dx + \int \frac{d - b}{(x + b)^2 + c - b^2} \, dx$$

Put $x = u - b$, then

$$I = \int \frac{u}{u^2 + c - b^2} \, du + (d - b) \int \frac{1}{u^2 + (c - b^2)} \, du \qquad (3.7.39)$$

The first integral is of the form

$$\int \frac{g'(u)}{g(u)} \, du$$

and can be evaluated using (3.7.13) as

$$\int \frac{u}{u^2 + c - b^2} \, du = \frac{1}{2} \ln |u^2 + (c - b^2)| \qquad\qquad (3.7.40)$$

The method of approach for the second integral of I is dictated by the nature of the constant $c - b^2$.

If $c - b^2 > 0$, write $c - b^2 = \alpha^2$, then from (3.7.8)

$$\int \frac{1}{u^2 + c - b^2} \, du = \int \frac{1}{u^2 + \alpha^2} \, du = \frac{1}{\alpha} \tan^{-1}(u/\alpha) \quad (3.7.41)$$

If $c - b^2 = 0$

$$\int \frac{1}{u^2 + c - b^2} \, du = \int \frac{du}{u^2} = -\frac{1}{u} \quad (3.7.42)$$

If $c - b^2 < 0$, write $c - b^2 = -\beta^2$, then from (3.7.34)

$$\int \frac{1}{u^2 + c - b^2} \, du = \int \frac{1}{u^2 - \beta^2} \, du = \frac{1}{2\beta} \ln \left| \frac{u - \beta}{u + \beta} \right| \quad (3.7.43)$$

Hence, returning to the original variable x the integral I can be expressed as follows.

If $c - b^2 > 0$

$$\int \frac{x + d}{x^2 + 2bx + c} \, dx = \frac{1}{2} \ln |x^2 + 2bx + c|$$

$$+ \frac{d - b}{(c - b^2)^{1/2}} \tan^{-1} \left\{ \frac{x + b}{(c - b^2)^{1/2}} \right\}$$

$$(3.7.44)$$

If $c - b^2 = 0$

$$\int \frac{x + d}{x^2 + 2bx + c} \, dx = \ln |x + b| + \frac{b - d}{x + b} \quad (3.7.45)$$

If $c - b^2 < 0$

$$\int \frac{x + d}{x^2 + 2bx + c} \, dx$$

$$= \frac{1}{2} \ln |x^2 + 2bx + c|$$

$$+ \frac{d - b}{2(b^2 - c)^{1/2}} \ln \left| \frac{x + b - (b^2 - c)^{1/2}}{x + b + (b^2 - c)^{1/2}} \right| \quad (3.7.46)$$

Trigonometric Functions

The following list of integrals involving trigonometric functions is not exhaustive but gives a guide to the techniques involved with this type of function.

(a) $\int \sin^2 x \, dx$ This integral is typical of many, which can be transformed to a more elementary form using the standard formulae for cosine and sine of multiple angles.

In this case

$$\cos 2x = 1 - 2 \sin^2 x \qquad (3.7.47)$$

so that the integral becomes

$$\int \sin^2 x \, dx = \frac{1}{2} \int 1 - \cos 2x \, dx$$

$$= \frac{1}{2} \left(x - \frac{1}{2} \sin 2x \right)$$

$$= \frac{1}{2} (x - \sin x \cos x) \qquad (3.7.48)$$

Similarly, using the relationship

$$\cos 2x = 2 \cos^2 x - 1 \qquad (3.7.49)$$

it can be seen that

$$\int \cos^2 x \, dx = \frac{1}{2} (x + \sin x \cos x) \qquad (3.7.50)$$

Alternatively

$$\int \cos^2 x \, dx = \int (1 - \sin^2 x) \, dx$$

$$= x - \frac{1}{2} (x - \sin x \cos x)$$

$$= \frac{1}{2} (x + \sin x \cos x)$$

(b) $\int \sin^3 x \, dx$ Using a similar technique to that used above, note that

$$\sin 3x = 3 \sin x - 4 \sin^3 x \qquad (3.7.51)$$

then

$$\int \sin^3 x \, dx = \frac{1}{4} \int (3 \sin x - \sin 3x) \, dx$$

$$= \frac{1}{4} \left(-3 \cos x + \frac{1}{3} \cos 3x \right) \qquad (3.7.52)$$

This can be simplified by use of the formula

$$\cos 3x = 4\cos^3 x - 3\cos x \tag{3.7.53}$$

so that the integral becomes

$$\int \sin^3 x \, dx = \frac{1}{3}\cos x(\cos^2 x - 3) \tag{3.7.54}$$

Similarly it may be shown that

$$\int \cos^3 x \, dx = \frac{1}{3}\sin x(3 - \sin^2 x) \tag{3.7.55}$$

In many instances the method of evaluation of an integral is not unique, and alternative approaches are perfectly valid. These integrals involving trigonometric functions are typical. For example $\int \sin^3 x \, dx$ can be evaluated using the technique of integration by parts as follows

$$\int \sin^3 x \, dx = \int \sin^2 x \sin x \, dx$$

$$= \sin^2 x(-\cos x) + \int \cos x \cdot \frac{d}{dx}(\sin^2 x) \, dx$$

$$= -\sin^2 x \cos x + 2 \int \cos x \sin x \cos x \, dx$$

$$= -\cos x \sin^2 x + 2 \int \sin x(1 - \sin^2 x) \, dx$$

$$= -\cos x(1 - \cos^2 x) + 2 \int \sin x \, dx$$

$$\qquad - 2 \int \sin^3 x \, dx$$

Hence

$$3 \int \sin^3 x \, dx = -\cos x(1 - \cos^2 x) - 2\cos x$$

so that

$$\int \sin^3 x \, dx = \frac{1}{3}\cos x(\cos^2 x - 3)$$

which agrees with (3.7.54).

(c) $\int \tan x \, dx$ Using the definition of $\tan x$

$$\int \tan x \, dx = \int \frac{\sin x}{\cos x} \, dx$$

$$= -\int \frac{1}{\cos x} \frac{d}{dx} \cos x \, dx$$

$$= -\ln|\cos x| \qquad \text{using (3.7.13)}$$

$$= \ln|\sec x| \qquad (3.7.56)$$

(d) $\int \sec x \, dx$ Many integrals involving trigonometric functions can be found by a simple substitution. This integral is a simple example of the technique.

Put $t = \tan \frac{1}{2} x$, then

$$\cos x = \frac{1 - t^2}{1 + t^2} \qquad \text{and} \qquad \sin x = \frac{2t}{1 + t^2} \qquad (3.7.57)$$

Also

$$\frac{dt}{dx} = \frac{1}{2} \sec^2 \frac{1}{2} x = \frac{1}{2}(1 + t^2) \qquad (3.7.58)$$

With this substitution

$$\int \sec x \, dx = \int \frac{1 + t^2}{1 - t^2} \cdot \frac{dx}{dt} \, dt$$

$$= \int \frac{1 + t^2}{1 - t^2} \cdot \frac{2}{1 + t^2} \cdot dt$$

$$= 2 \int \frac{dt}{1 - t^2}$$

$$= \ln \left| \frac{1 + t}{1 - t} \right| \qquad \text{from (3.7.34)}$$

Therefore recovering the integral as a function of x

$$\int \sec x \, dx = \ln \left| \frac{1 + \tan \frac{1}{2} x}{1 - \tan \frac{1}{2} x} \right| \qquad (3.7.59)$$

$$= \ln \left| \tan \left(\frac{\pi}{4} + \frac{1}{2} x \right) \right| \qquad (3.7.60)$$

This result can be put in an alternative form which may prove more useful in some instances.

From (3.7.59)

$$\int \sec x \, dx = \ln \left| \frac{1 + \tan \frac{1}{2}x}{1 - \tan \frac{1}{2}x} \right|$$

$$= \ln \left| \frac{\cos \frac{1}{2}x + \sin \frac{1}{2}x}{\cos \frac{1}{2}x - \sin \frac{1}{2}x} \right|$$

$$= \ln \left| \frac{(\cos \frac{1}{2}x + \sin \frac{1}{2}x)}{(\cos \frac{1}{2}x - \sin \frac{1}{2}x)} \cdot \frac{(\cos \frac{1}{2}x + \sin \frac{1}{2}x)}{(\cos \frac{1}{2}x + \sin \frac{1}{2}x)} \right|$$

$$= \ln \left| \frac{\cos^2 \frac{1}{2}x + 2 \sin \frac{1}{2}x \cos \frac{1}{2}x + \sin^2 \frac{1}{2}x}{\cos^2 \frac{1}{2}x - \sin^2 \frac{1}{2}x} \right|$$

$$= \ln \left| \frac{1 + \sin x}{\cos x} \right|$$

Alternatively

$$\int \sec x \, dx = \ln |\sec x + \tan x| \qquad (3.7.61)$$

(e) $\int \sec^3 x \, dx$ This again is a typical example of an integral which is found by integration by parts as

$$\int \sec^3 x \, dx = \int \sec x \cdot \sec^2 x \, dx$$

$$= \int \sec x \frac{d}{dx} (\tan x) \, dx$$

$$= \sec x \tan x - \int \sec x \tan x \cdot \tan x \, dx$$

$$= \sec x \tan x - \int \sec x (\sec^2 x - 1) \, dx$$

$$= \sec x \tan x - \int \sec^3 x \, dx + \int \sec x \, dx$$

Hence

$$2 \int \sec^3 x \, dx = \sec x \tan x + \ln |\sec x + \tan x|$$

giving

$$\int \sec^3 x \, dx = \frac{1}{2} \{ \sec x \tan x + \ln |\sec x + \tan x| \} \qquad (3.7.62)$$

81

There are many more methods available for more complicated integrands but in order to include all of them the text would need to be considerably larger or even devoted entirely to the subject of integration. However the techniques demonstrated above should give the reader a basic understanding of the process of integration, and allow him to proceed to more demanding texts.

3.8 Approximate Methods for Evaluating Definite Integrals

In many practical examples it is impossible to evaluate a definite integral by means of simple well tabulated functions. This may be either because the integrand is a complicated function and no simple antiderivative can be found or that, even though the integrand is well defined at every point it can only be found by observation and it would only be practical to make observations at a finite number of points.

When this situation arises approximate numerical techniques have to be employed. There are many such approximations but two of the most widely used are the Trapezoidal Rule and Simpson's Rule.

Trapezoidal Rule

The integral under consideration is

$$\int_a^b f(x)\,dx$$

which is equal to the area under the curve $y = f(x)$ and bounded by the ordinates $x = a$ and $x = b$ (see Figure 3.3).

Let the area be divided into N strips each of width h by the ordinates

$$x_n = a + nh \qquad n = 1, 2, 3, \ldots, N - 1 \tag{3.8.1}$$

where $h = (b - a)/N$.

At these points the values of $y = f(x)$ are

$$y_n = f(a + nh) \qquad n = 1, 2, 3, \ldots, N - 1 \tag{3.8.2}$$

and

$$y_0 = f(a) \qquad y_N = f(b) \tag{3.8.3}$$

The area S_n contained by a typical strip may be approximated by the area of the trapezium $P_n A_n A_{n+1} P_{n+1}$ as

$$S_n = h \cdot \tfrac{1}{2}(y_n + y_{n+1}) \tag{3.8.4}$$

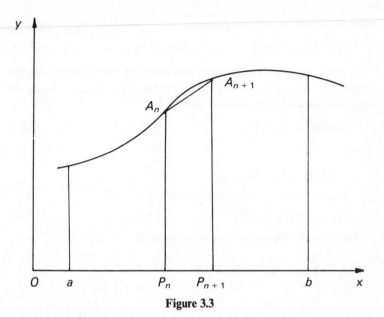

Figure 3.3

The total area under the curve can now be found as the sum of the areas contained by the strips, so that

$$\int_a^b f(x)\,dx \doteq \sum_{n=0}^{N-1} S_n$$

$$= \frac{1}{2} h \sum_{n=0}^{N-1} (y_n + y_{n+1})$$

Hence

$$\int_a^b f(x)\,dx = \frac{1}{2} h \left\{ y_0 + y_N + 2 \sum_{n=1}^{N-1} y_n \right\} \tag{3.8.5}$$

Simpson's Rule

In this case let the area under the curve be divided into an even number of strips each of width h by ordinates

$$x_n = a + nh \qquad n = 1, 2, 3, \ldots, 2N - 1 \tag{3.8.6}$$

where $h = (b - a)/2N$.

At these points the function takes the values

$$y_n = f(a + nh) \qquad n = 1, 2, 3, \ldots, 2N - 1 \tag{3.8.7}$$

83

with

$$y_0 = f(a) \qquad \text{and} \qquad y_{2N} = f(b) \tag{3.8.8}$$

In order to calculate the area contained by two typical adjacent strips let the curve through the points A_{2n}, A_{2n+1}, A_{2n+2} be approximated by a parabola (Figure 3.4)

$$y = \alpha x^2 + \beta x + \gamma \tag{3.8.9}$$

Then

$$
\begin{aligned}
y_{2n} &= \alpha x_{2n}^2 + \beta x_{2n} + \gamma \\
y_{2n+1} &= \alpha(x_{2n} + h)^2 + \beta(x_{2n} + h) + \gamma \\
y_{2n+2} &= \alpha(x_{2n} + 2h)^2 + \beta(x_{2n} + 2h) + \gamma
\end{aligned}
\tag{3.8.10}
$$

The area contained by the pair of strips is

$$
\begin{aligned}
S_n &= \int_{x_{2n}}^{x_{2n}+2h} \alpha x^2 + \beta x + \gamma \, dx \\
&= \left[\frac{\alpha x^3}{3} + \frac{\beta x^2}{2} + \gamma x \right]_{x_{2n}}^{x_{2n}+2h} \\
&= \tfrac{1}{3}h\{2\alpha(3x_{2n}^2 + 6hx_{2n} + 4h^2) + 6\beta(x_{2n} + h) + 6\gamma\} \\
&= \tfrac{1}{3}h\{y_{2n} + 4y_{2n+1} + y_{2n+2}\}
\end{aligned}
\tag{3.8.11}
$$

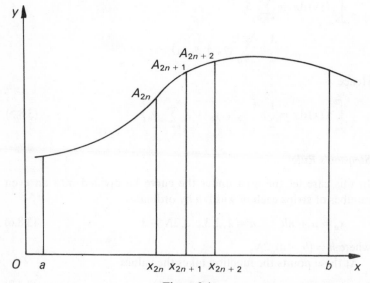

Figure 3.4

Again the total area under the curve is found approximately as the sum of the areas S_n. Thus the integral is

$$\int_a^b f(x)\,dx \doteq \sum_{n=0}^{N-1} S_n$$

$$= \frac{1}{3} h \sum_{n=0}^{N-1} (y_{2n} + 4y_{2n+1} + y_{2n+2})$$

Hence

$$\int_a^b f(x)\,dx \doteq \frac{1}{3} h \left\{ y_0 + y_N + 2 \sum_{n=1}^{N-1} y_{2n} + 4 \sum_{n=0}^{N-1} y_{2n+1} \right\} \quad (3.8.12)$$

The accuracy of both of these approximate formulae is dependent on the size of h, the strip width, or the number of strips used. It can be shown that the largest possible errors are $M_2 h^2 (b - a)/12$, for the Trapezoidal rule and $M_4 h^4 (b - a)/90$, for Simpson's rule, where M_2 and M_4 are the largest numerical values of $f''(x)$ and $f^4(x)$, respectively.

Example By using both the Trapezoidal rule and Simpson's rule with 5 ordinates find an approximate value for

$$\int_0^1 \frac{dx}{1 + x^2}$$

With the above notation $a = 0$, $b = 1$ and $h = 0.25$. Then

$$y_n = \frac{1}{1 + x_n^2} \quad (3.8.13)$$

where $x_n = 0.25n$, $\quad n = 0, 1, 2, 3, 4$.

The value of the ordinates are displayed in Table 3.1.

Table 3.1

n	0	1	2	3	4
x_n	0	0.25	0.5	0.75	1.0
y_n	1	0.9411765	0.8	0.64	0.5

For the Trapezoidal rule, $N = 4$ and

$$\sum_{n=1}^{N-1} y_n = \sum_{n=1}^{3} y_n = y_1 + y_2 + y_3 = 2.3811765 \quad (3.8.14)$$

85

Hence using (3.8.5)

$$\int_0^1 \frac{dx}{1+x^2} = \frac{1}{2} \times 0.25 \times \{1 + 0.5 + 2 \times 2.3811765\}$$

$$= 0.7827941 \qquad (3.8.15)$$

For Simpson's rule, $N = 2$ and

$$\sum_{n=1}^{N-1} y_{2n} = y_2 = 0.8 \qquad (3.8.16)$$

$$\sum_{n=0}^{N-1} y_{2n+1} = \sum_{n=0}^{1} y_{2n+1} = y_1 + y_3 = 1.5811765 \qquad (3.8.17)$$

Hence using (3.8.12)

$$\int_0^1 \frac{dx}{1+x^2} = \frac{1}{3} \times 0.25 \times \{1 + 0.5 + 2 \times 0.8 + 4 \times 1.5811765\}$$

$$= 0.7853922 \qquad (3.8.18)$$

In this particular example the integral can be evaluated exactly as

$$\int_0^1 \frac{dx}{1+x^2} = [\tan^{-1} x]_0^1 = \frac{\pi}{4} = 0.7853982 \qquad (3.8.19)$$

Thus, in this case, even with a relatively small number of subdivisions there is good agreement between the approximate and exact values for the integral. As anticipated, Simpson's rule gives the better approximation to the true value.

3.9 Summary of Simple Standard Integrals

$f(x)$	$\int f(x)\,dx$
$x^n, n \neq -1$	$x^{n+1}/(n+1)$
$1/x$	$\ln|x|$
e^{ax}	$\dfrac{1}{a} e^{ax}$
$\cos ax$	$\dfrac{1}{a} \sin ax$

$f(x)$	$\int f(x)\,\mathrm{d}x$
$\sin ax$	$-\dfrac{1}{a}\cos ax$
$\tan ax$	$-\dfrac{1}{a}\ln|\cos ax|$
$\sec^2 ax$	$\dfrac{1}{a}\tan ax$
$\operatorname{cosec}^2 ax$	$-\dfrac{1}{a}\cot ax$
$\sec ax \tan ax$	$\dfrac{1}{a}\sec ax$
$\dfrac{1}{(a^2 - x^2)^{1/2}}$	$\sin^{-1}\dfrac{x}{a}$
$\dfrac{1}{a^2 + x^2}$	$\dfrac{1}{a}\tan^{-1}\dfrac{x}{a}$
$\dfrac{1}{a^2 - x^2}$	$\dfrac{1}{2a}\ln\left|\dfrac{x+a}{x-a}\right|$
$\cosh ax$	$\dfrac{1}{a}\sinh ax$
$\sinh ax$	$\dfrac{1}{a}\cosh ax$
$\operatorname{sech}^2 ax$	$\dfrac{1}{a}\tanh ax$
$\operatorname{cosech}^2 ax$	$\dfrac{1}{a}\coth ax$
$\dfrac{1}{(a^2 + x^2)^{1/2}}$	$\sinh^{-1}\dfrac{x}{a}$
$\dfrac{1}{(x^2 - a^2)^{1/2}}$	$\cosh^{-1}\dfrac{x}{a}$
$\mathrm{e}^{ax}\sin bx$	$\mathrm{e}^{ax}[a\sin bx - b\cos bx]/(a^2 + b^2)$
$\mathrm{e}^{ax}\cos bx$	$\mathrm{e}^{ax}[a\cos bx + b\sin bx]/(a^2 + b^2)$

Exercises

3.1 Find the following indefinite integrals

(i) $\displaystyle\int \left(x + \frac{1}{x}\right)^2 dx$

(ii) $\displaystyle\int \frac{(\sqrt{x} - 1)^4}{x} dx$

(iii) $\displaystyle\int (x + 2)(x^2 + 3x - 1) dx$

3.2 By expressing the integrands as partial fractions find

(i) $\displaystyle\int \frac{dx}{(1 + x)(1 + x^2)}$

(ii) $\displaystyle\int \frac{dx}{(x - 2)(x + 1)}$

(iii) $\displaystyle\int \frac{2x^2 - 1}{x^2(x - 1)} dx$

(iv) $\displaystyle\int \frac{dx}{(x + 1)^2(x^2 + 1)}$

3.3 Using the method of integration by parts find

(i) $\displaystyle\int x \sin 2x \, dx$

(ii) $\displaystyle\int x \ln x \, dx$

(iii) $\displaystyle\int \ln(1 + x^2) \, dx$

(iv) $\displaystyle\int e^{4x} \sin 3x \, dx$

3.4 Find the integrals

(i) $\displaystyle\int \frac{\sin \theta}{(3 + \cos \theta)^{1/2}} d\theta$

(ii) $\displaystyle\int (1 + 2 \sin \theta)^3 \, d\theta$

3.5 By means of the substitution $t = \tan\frac{1}{2}x$ find

(i) $\displaystyle\int \frac{2 \sin x}{\sin x - \cos x} dx$

(ii) $\displaystyle\int \frac{dx}{\sin x(1 + \sin x)}$

3.6 Using the given substitution find the following integrals

(i) $\displaystyle\int \frac{x^2}{(1 - x^2)^{1/2}} dx$ $x = \sin \theta$

(ii) $\displaystyle\int \frac{2x + 3}{(1 - x^2)^{1/2}} dx$ $x = \sin \theta$

(iii) $\displaystyle\int_1^2 \frac{dx}{x^2(x^2 - 1)^{1/2}}$ $x = \sec \theta$

(iv) $\displaystyle\int_2^7 \frac{dx}{(x + 1)(x + 2)^{1/2}}$ $x = u^2 - 2$

3.7 Find the integrals

(i) $\displaystyle\int \frac{\sin t}{5 + \tan^2 t}\, dt$ (ii) $\displaystyle\int \tan^4 t\, dt$

3.8 Evaluate the definite integrals

(i) $\displaystyle\int_0^1 x(1 - x)^3\, dx$ (ii) $\displaystyle\int_{-1}^1 (x + 3)$
$\times (2x^2 + x - 1)\, dx$

(iii) $\displaystyle\int_0^{\pi/3} x^3 \sin 3x\, dx$ (iv) $\displaystyle\int_0^{\pi/2} \sin 3x \cos x\, dx$

3.9 Evaluate

(i) $\displaystyle\int_0^2 \sqrt{(3x + 2)}\, dx$ (ii) $\displaystyle\int_1^2 \frac{dx}{x(1 + x)^2}$

(iii) $\displaystyle\int_1^4 \frac{dx}{\sqrt{(8 + 2x - x^2)}}$.

3.10 Show that the substitution $x = \cos 2\theta$, reduces the integral

$$\int_0^1 \sqrt{\left(\frac{1 - x}{1 + x}\right)}\, dx$$

to the form

$$4 \int_0^{\pi/4} \sin^2 \theta\, d\theta$$

and hence evaluate the integral.

3.11 By the use of integration by parts prove the following recurrence relationships

(i) If $I_n = \displaystyle\int_0^{\pi/2} \sin^n x\, dx$ then $nI_n = (n - 1)I_{n-2}$

(ii) If $I_n = \displaystyle\int_0^{\pi/2} x \sin^n x\, dx$ then $n^2 I_n = n(n - 1)I_{n-2} + 1$

(iii) If $I_n = \displaystyle\int_0^a \frac{x^n}{(a^2 - x^2)^{1/2}}\, dx$ then $nI_n = (n - 1)a^2 I_{n-2}$

In each case evaluate I_4.

3.12 Find the area under the curve $y = x^2 + 2x$ between the points $x = 0$ and $x = 3$.

3.13 Find the area of the region bounded by the curves

$$xy = a^2 \qquad xy = 2a^2 \qquad y^2 = ax \qquad y^2 = 2ax$$

3.14 Find the area lying inside the circle $x^2 + y^2 - 2x = 0$ and above the parabola $y = x^2$.

3.15 Find the area of the loop of the curve $y^2 = x^2 - x^3$.

4
Partial Differentiation

4.1 Introduction

The concepts introduced in the previous chapters for functions of a single variable are extended and modified in this chapter to include functions of many variables.

For a function of one variable the ideas of limits and differentiation were introduced intuitively and then the definitions were formalised. In this chapter the formal approach is adopted immediately to avoid errors that sometimes result from a more intuitive approach when dealing with a function of several variables.

The ideas are discussed and defined in detail for a function of two variables, but are easily extended to cover a function of more than two variables.

4.2 Functions of Two Variables

DEFINITION   A quantity z is said to be a function of two variables x and y, if every pair of values (x, y) defines a unique value of z. In this case, z is written as

$$z = f(x, y) \tag{4.2.1}$$

or very often, for convenience

$$z = z(x, y) \tag{4.2.2}$$

For example

$$z = x^2 + y^2 \qquad z = \sin(x + y) \qquad z = \frac{x^3 + y^3}{xy} \tag{4.2.3}$$

are typical functions of two variables.

The variables x and y are said to be the independent variables and z is the dependent variable.

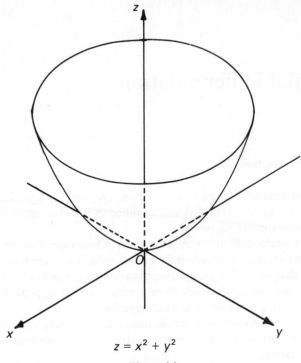

$$z = x^2 + y^2$$

Figure 4.1

Clearly the definition is readily extended to functions of more than two variables.

In the same way that a function $f(x)$ of a single variable can be represented as a curve $y = f(x)$ in two dimensions, a function of two variables can be represented as a surface

$$z = f(x, y)$$

in three dimensional cartesian space. A typical example is shown in Figure 4.1.

This geometrical interpretation does not extend usefully to functions of more than two variables.

4.3 Limits

The concept of a limit can be extended through a formal definition to functions of two variables.

92

DEFINITION   A function $f(x, y)$ has a limit L at the point (a, b) if given any small value $\varepsilon > 0$, there exists a value $\delta > 0$ so that for all x, y such that

$$(x - a)^2 + (y - b)^2 < \delta^2 \tag{4.3.1}$$

then

$$|f(x, y) - L| < \varepsilon \tag{4.3.2}$$

This limit is written as

$$L = \lim_{\substack{x \to a \\ y \to b}} f(x, y) \tag{4.3.3}$$

or

$$L = \lim_{(x,y) \to (a,b)} f(x, y) \tag{4.3.4}$$

The definition can be thought of in geometrical terms as follows.

A function $f(x, y)$ has a limit L at the point (a, b) if given any number $\varepsilon > 0$ it is possible to construct a circle centre (a, b) so that

$$|f(x, y) - L| < \varepsilon$$

for every point (x, y) in that circle.

In the terms already introduced in Chapter 1, the value of the function $f(x, y)$ 'tends to L' as the point (x, y) approaches the point (a, b). The value of L must be independent of the manner in which point (a, b) is approached.

An example of a function whose behaviour does depend on the manner of approach is given in the following example.

Example   Show that the function

$$f(x, y) = y/x$$

does not possess a limit as (x, y) approaches the origin $(0, 0)$.

Suppose that (x, y) approaches $(0, 0)$ but that the approach is constrained so that the point always lies on the line $y = mx$.

If (x, y) lies on $y = mx$

$$f(x, y) = \frac{y}{x} = \frac{mx}{x} = m \tag{4.3.5}$$

Thus, as (x, y) moves closer and closer to $(0, 0)$ the function always keeps the constant value m. The value of m depends on the chosen line of approach and the limit of $f(x, y)$ will therefore depend on the way in which (x, y) is allowed to approach the point $(0, 0)$.

93

An approach along $y = 0$ gives a limit equal to zero, while an approach along $y = x$ gives a value 1.

Hence the function $f(x, y) = y/x$ has no limit at the point $(0, 0)$.

It is clear from this example that, for a function of two variables, the way in which limits are taken is important. In particular, the order in which limits are taken can be critical.

The limits

$$\lim_{x \to a} \lim_{y \to b} f(x, y) \qquad \text{and} \qquad \lim_{y \to b} \lim_{x \to a} f(x, y) \qquad (4.3.6)$$

may both exist but take different values. In such a case, the limit

$$\lim_{\substack{x \to a \\ y \to b}} f(x, y)$$

does not exist.

Example Consider the behaviour of $f(x, y) = (x^2 - y^2)/(x^2 + y^2)$ as (x, y) approaches the origin $(0, 0)$.

If x is fixed and y allowed to approach zero the following limit results

$$\lim_{y \to 0} f(x, y) = \lim_{y \to 0} \frac{x^2 - y^2}{x^2 + y^2} = \frac{x^2}{x^2} = 1 \qquad (4.3.7)$$

Now, allowing x to tend to zero

$$\lim_{x \to 0} \lim_{y \to 0} f(x, y) = \lim_{x \to 0} (1) = 1 \qquad (4.3.8)$$

Similarly, taking the limits in the opposite order

$$\lim_{x \to 0} f(x, y) = \lim_{x \to 0} \frac{x^2 - y^2}{x^2 + y^2} = -\frac{y^2}{y^2} = -1 \qquad (4.3.9)$$

and

$$\lim_{y \to 0} \lim_{x \to 0} f(x, y) = \lim_{y \to 0} (-1) = -1 \qquad (4.3.10)$$

Therefore, for this particular function

$$\lim_{x \to 0} \lim_{y \to 0} f(x, y) \neq \lim_{y \to 0} \lim_{x \to 0} f(x, y) \qquad (4.3.11)$$

In this example the order in which limits are taken is important. Since the value of the limit is different when the order of taking limits is reversed, $f(x, y)$ is said to have no limit at the point $(0, 0)$.

94

DEFINITION A function $f(x, y)$ is said to be continuous at the point (a, b), if

$$\lim_{\substack{x \to a \\ y \to b}} f(x, y)$$

exists and is equal to $f(a, b)$, the value of the function at (a, b).

4.4 Partial Derivatives

A function $f(x, y)$ takes on different values as different values are assigned to the independent variables x and y. In order to describe the variations of the function $f(x, y)$ as a result of the variations in x and y the concept of partial differentiation is introduced.

In the case of a function of a single variable the derivative measures the rate of change of the function with respect to that variable. For a function of two variables, to describe any variation completely, two partial derivatives are necessary to measure the rate of change with respect to each of the two independent variables in turn.

DEFINITION If $f(x, y)$ is a function of two variables (x, y) then provided that the limit exists, the partial derivative of $f(x, y)$, with respect to x, at the point (a, b) is defined as

$$f_x(a, b) = \lim_{h \to 0} \frac{f(a + h, b) - f(a, b)}{h} \qquad (4.4.1)$$

Similarly

$$f_y(a, b) = \lim_{k \to 0} \frac{f(a, b + k) - f(a, b)}{k} \qquad (4.4.2)$$

is the partial derivative of $f(x, y)$ with respect to y at the point (a, b).

The process of obtaining these partial derivatives is called partial differentiation.

In general, partial differentiation can be thought of as differentiation with respect to one of the variables while the second variable is kept fixed, that is differentiation with respect to x along a line $y = $ constant, or differentiation with respect to y along a line $x = $ constant. There are however cases where care is needed in applying this method. Indeed, for some functions, the process may break down at certain points.

When a function has a discontinuity at a point (a, b) difficulties may arise. The derivatives $f_x(a, b)$ and $f_y(a, b)$ may exist and in that case can be found by application of the formal definition. However, it is not possible to obtain these derivatives by differentiating the function in turn with respect to x and y and then setting $x = a$ and $y = b$.

Example Show that the function $f(x, y)$ defined by

$$f(x, y) = \frac{xy}{(x^2 + y^2)} \qquad (x, y) \neq (0, 0)$$

$$f(0, 0) = 0$$

is discontinuous at $(0, 0)$, but the derivatives $f_x(0, 0)$ and $f_y(0, 0)$ exist and are both equal to zero.

Let (x, y) approach $(0, 0)$ along the line $y = mx$, then

$$f(x, y) = \frac{xy}{x^2 + y^2} = \frac{mx^2}{x^2 + m^2 x^2} = \frac{m}{1 + m^2} \qquad (4.4.3)$$

so that the value of f depends on the choice of line. In these circumstances f has no limit as (x, y) approaches $(0, 0)$ since its value depends on the manner of approach. If $f(0, 0)$ is the value of f at $(0, 0)$, it is clear that

$$f(0, 0) \neq \lim_{\substack{x \to 0 \\ y \to 0}} f(x, y) \qquad (4.4.4)$$

because the limit does not exist. Therefore the function cannot be continuous at $(0, 0)$.

The formal definition of the partial derivative gives

$$f_x(0, 0) = \lim_{h \to 0} \left\{ \frac{f(h, 0) - f(0, 0)}{h} \right\}$$

$$= \lim_{h \to 0} \left\{ \frac{0 - 0}{h} \right\}$$

$$= \lim_{h \to 0} 0$$

$$= 0 \qquad (4.4.5)$$

By symmetry it follows that

$$f_y(0, 0) = 0 \qquad (4.4.6)$$

It can now be demonstrated that these results do not follow by setting $x = 0$, $y = 0$ in general expressions obtained for $f_x(x, y)$ and $f_y(x, y)$.

96

If y is held fixed the function $f(x, y)$ can be differentiated, as a function of the single variable x, to give

$$f_x(x, y) = \frac{(x^2 + y^2)y - xy \cdot 2x}{(x^2 + y^2)^2}$$

$$= -\frac{(x^2 - y^2)y}{(x^2 + y^2)^2} \qquad (4.4.7)$$

Similarly, treating x as a constant and differentiating with respect to y, the partial derivative $f_y(x, y)$ is given as

$$f_y(x, y) = \frac{(x^2 - y^2)x}{(x^2 + y^2)^2} \qquad (4.4.8)$$

It is not now valid to set $x = 0$, $y = 0$ to obtain $f_x(0,0) f_y(0,0)$ since neither of the expressions (4.4.7) or (4.4.8) has a limit as $(x, y) \to (0, 0)$. Therefore, although the derivatives are known to exist they cannot be calculated by the simple mechanical process of differentiation.

This example clearly illustrates that when a function has points of discontinuity care is required in calculating derivatives at those points.

It is also seen from this example, that, unlike functions of a single variable, existence of the partial derivatives of a function of two variables does not imply continuity of that function.

If the functions are well behaved and continuous everywhere there are no problems, and the derivatives can be simply obtained by routine differentiation.

As in the case of ordinary differentiation there are various notations in use for partial derivatives. A popular alternative to f_x and f_y is

$$\frac{\partial f}{\partial x} \qquad \frac{\partial f}{\partial y} \qquad (4.4.9)$$

to denote the derivatives with respect to x and y, respectively.

It is occasionally necessary to emphasize the fact that one of the variables is held constant. In that case the notation

$$\left(\frac{\partial f}{\partial x}\right)_y \qquad \left(\frac{\partial f}{\partial y}\right)_x \qquad (4.4.10)$$

is often employed. The former denotes differentiation with respect to x emphasizing that y is held constant, and by symmetry the latter denotes differentiation with respect to y, keeping x fixed.

4.5 Higher Order Partial Derivatives

The partial derivatives $f_x(x, y)$ and $f_y(x, y)$ are themselves functions of x and y and may therefore possess derivatives with respect to both x and y.

These derivatives are usually written as

$$f_{xx}(x, y) \qquad f_{xy}(x, y) \qquad f_{yx}(x, y) \qquad f_{yy}(x, y) \qquad (4.5.1)$$

The derivative $f_{xy}(x, y)$ for example denotes the derivative of $f_x(x, y)$ with respect to y. These expressions are called the second derivatives or partial derivatives of order two.

It can be shown that, if the functions $f_{xy}(x, y)$ and $f_{yx}(x, y)$ are both continuous, then they are equal.

In the notation of (4.4.9) these second derivatives are written as

$$\frac{\partial^2 f}{\partial x^2} \qquad \frac{\partial^2 f}{\partial y \partial x} \qquad \frac{\partial^2 f}{\partial x \partial y} \qquad \frac{\partial^2 f}{\partial y^2} \qquad (4.5.2)$$

With this notation

$$\frac{\partial^2 f}{\partial x^2} = \frac{\partial}{\partial x}\left(\frac{\partial f}{\partial x}\right) \qquad \frac{\partial^2 f}{\partial y \partial x} = \frac{\partial}{\partial y}\left(\frac{\partial f}{\partial x}\right) \qquad \text{etc.} \qquad (4.5.3)$$

The notation clearly indicates the order in which the repeated differentiations are performed.

4.6 Differentials

If $f(x, y)$ has continuous partial derivatives in a region R of the (x, y) plane then if (a, b) lies within R

$$f(a + h, b + k) = f(a, b) + hf_x(a, b) + kf_y(a, b) + \varepsilon_1 h + \varepsilon_2 k$$

where $\varepsilon_1 \to 0$ and $\varepsilon_2 \to 0$ as $h, k \to 0$.

PROOF   Let the increment in f when x is changed from a to $a + h$ and y is changed from b to $b + k$, be Δf so that

$$\Delta f = f(a + h, b + k) - f(a, b) \qquad (4.6.1)$$

This can be rewritten as

$$\Delta f = f(a + h, b + k) - f(a, b + k) + f(a, b + k) - f(a, b)$$

$$(4.6.2)$$

The first two terms give the increment obtained by keeping the second variable fixed at the value $b + k$ and changing the first from a to $a + h$.

Using the mean value Theorem (2.8.3) these two terms can be combined as

$$f(a + h, b + k) - f(a, b + k) = hf_x(a + \theta_1 h, b + k) \qquad (4.6.3)$$

where $0 < \theta_1 < 1$.

Similarly the last two terms of (4.6.2) combine to give

$$f(a, b + k) - f(a, b) = kf_y(a, b + \theta_2 k) \qquad (4.6.4)$$

where $0 < \theta_2 < 1$.

Now, since $f_x(x, y)$ and $f_y(x, y)$ are continuous in the region R it follows that

$$f_x(a + \theta_1 h, b + k) = f_x(a, b) + \varepsilon_1 \qquad (4.6.5)$$

and

$$f_y(a, b + \theta_2 k) = f_y(a, b) + \varepsilon_2 \qquad (4.6.6)$$

where ε_1 and ε_2 both tend to zero as h and k both tend to zero.

Therefore, substitution back from (4.6.3), (4.6.4), (4.6.5) and (4.6.6) into (4.6.2) gives

$$\Delta f = h\{f_x(a, b) + \varepsilon_1\} + k\{f_y(a, b) + \varepsilon_2\} \qquad (4.6.7)$$

that is

$$f(a + h, b + k) = f(a, b) + hf_x(a, b) + kf_y(a, b) + h\varepsilon_1 + k\varepsilon_2 \qquad (4.6.8)$$

where $\varepsilon_1 \to 0$ and $\varepsilon_2 \to 0$ as $h, k \to 0$.

Using the notation of small increments and writing $h = \mathrm{d}x$, $k = \mathrm{d}y$ the increment

$$\mathrm{d}f = f_x(a, b)\,\mathrm{d}x + f_y(a, b)\,\mathrm{d}y \qquad (4.6.9)$$

is called the differential of f or the principal part of Δf.

DEFINITION   If a function $f(x, y)$ satisfies the condition (4.6.8) then $f(x, y)$ is said to be differentiable at the point (a, b).

From above, it is seen that a function with continuous partial derivatives is necessarily differentiable.

4.7 Chain Rules

(a) If $z = f(x, y)$ is a function with continuous partial derivatives and x, y can be expressed in terms of two other variables u, v by $x = x(u, v)$, $y = y(u, v)$, then

$$\frac{\partial z}{\partial u} = \frac{\partial f}{\partial x} \cdot \frac{\partial x}{\partial u} + \frac{\partial f}{\partial y} \cdot \frac{\partial y}{\partial u}$$

$$\frac{\partial z}{\partial v} = \frac{\partial f}{\partial x} \cdot \frac{\partial x}{\partial v} + \frac{\partial f}{\partial y} \cdot \frac{\partial y}{\partial v} \tag{4.7.1}$$

In general if $z = f(x_1, x_2, \ldots, x_n)$ is a function of n variables and $x_r = x_r(u_1, u_2, \ldots, u_n)$, $r = 1, 2, \ldots, n$ then

$$\frac{\partial z}{\partial u_r} = \frac{\partial f}{\partial x_1} \cdot \frac{\partial x_1}{\partial u_r} + \frac{\partial f}{\partial x_2} \cdot \frac{\partial x_2}{\partial u_r} + \cdots + \frac{\partial f}{\partial x_n} \cdot \frac{\partial x_n}{\partial u_r}$$

$$= \sum_{i=1}^{n} \frac{\partial f}{\partial x_i} \frac{\partial x_i}{\partial u_r} \qquad r = 1, 2, \ldots, n \tag{4.7.2}$$

PROOF   Let a small change Δu in u, yield changes $\Delta x, \Delta y$ and Δz in x, y and z, respectively, then

$$\frac{\partial z}{\partial u} = \lim_{\Delta u \to 0} \frac{\Delta z}{\Delta u} \tag{4.7.3}$$

Assuming that $f(x, y)$, $x(u, v)$ and $y(u, v)$ are differentiable it follows from (4.6.8) that

$$\Delta z = \Delta x \frac{\partial f}{\partial x} + \Delta y \frac{\partial f}{\partial y} + \Delta x \varepsilon_1 + \Delta y \varepsilon_2 \tag{4.7.4}$$

where $\varepsilon_1 \to 0$ and $\varepsilon_2 \to 0$ as $\Delta u \to 0$.

Therefore, combining (4.7.3) and (4.7.4) it follows that

$$\frac{\partial z}{\partial u} = \lim_{\Delta u \to 0} \left\{ \frac{\Delta x}{\Delta u} \frac{\partial f}{\partial x} + \frac{\Delta y}{\Delta u} \frac{\partial f}{\partial y} + \frac{\Delta x}{\Delta u} \varepsilon_1 + \frac{\Delta y}{\Delta u} \varepsilon_2 \right\}$$

$$= \frac{\partial f}{\partial x} \cdot \frac{\partial x}{\partial u} + \frac{\partial f}{\partial y} \cdot \frac{\partial y}{\partial u} + \frac{\partial x}{\partial u} \cdot \lim_{\Delta u \to 0} \varepsilon_1 + \frac{\partial y}{\partial u} \lim_{\Delta u \to 0} \varepsilon_2$$

That is

$$\frac{\partial z}{\partial u} = \frac{\partial f}{\partial x} \cdot \frac{\partial x}{\partial u} + \frac{\partial f}{\partial y} \cdot \frac{\partial y}{\partial u} \tag{4.7.5}$$

Similarly, it can be shown that

$$\frac{\partial z}{\partial v} = \frac{\partial f}{\partial x} \cdot \frac{\partial x}{\partial v} + \frac{\partial f}{\partial y} \cdot \frac{\partial y}{\partial v} \qquad (4.7.6)$$

(b) If $z = f(x, y)$ is a function with continuous partial derivatives and x, y can be written in terms of a single parameter t as $x = x(t)$, $y = y(t)$, then

$$\frac{dz}{dt} = \frac{\partial f}{\partial x} \cdot \frac{dx}{dt} + \frac{\partial f}{\partial y} \cdot \frac{dy}{dt} \qquad (4.7.7)$$

PROOF   The proof of this proposition follows exactly the same lines as the previous proof.

Let a change Δt lead to increments Δx, Δy and Δz, respectively. Then

$$\frac{dz}{dt} = \lim_{\Delta t \to 0} \frac{\Delta z}{\Delta t}$$

$$= \lim_{\Delta t \to 0} \frac{1}{\Delta t} \left\{ \frac{\partial f}{\partial x} \Delta x + \frac{\partial f}{\partial y} \Delta y + \Delta x \varepsilon_1 + \Delta y \varepsilon_2 \right\} \qquad (4.7.8)$$

where $\varepsilon_1 \to 0$ and $\varepsilon_2 \to 0$ as $\Delta t \to 0$.

Therefore

$$\frac{dz}{dt} = \lim_{\Delta t \to 0} \left\{ \frac{\partial f}{\partial x} \frac{\Delta x}{\Delta t} + \frac{\partial f}{\partial y} \cdot \frac{\Delta y}{\Delta t} + \varepsilon_1 \frac{\Delta x}{\Delta t} + \varepsilon_2 \frac{\Delta y}{\Delta t} \right\}$$

$$= \frac{\partial f}{\partial x} \cdot \frac{dx}{dt} + \frac{\partial f}{\partial y} \cdot \frac{dy}{dt} \qquad (4.7.9)$$

A particular consequence of this result is the formula for dy/dx when y is implicitly related to x by the implicit function $f(x, y) = 0$.

Differentiating $f(x, y) = 0$ with respect to x remembering that y is a function of x, it follows from (4.7.9) that

$$0 = \frac{\partial f}{\partial x} \cdot \frac{dx}{dx} + \frac{\partial f}{\partial y} \cdot \frac{dy}{dx}$$

$$= \frac{\partial f}{\partial x} + \frac{\partial f}{\partial y} \cdot \frac{dy}{dx} \qquad (4.7.10)$$

Hence

$$\frac{dy}{dx} = -\frac{\partial f}{\partial x} \bigg/ \frac{\partial f}{\partial y} \qquad (4.7.11)$$

101

Examples

(i) If $x = r \cos \theta$, $y = r \sin \theta$ find $\dfrac{\partial z}{\partial x}$ and $\dfrac{\partial z}{\partial y}$ in terms of $\dfrac{\partial z}{\partial r}$ and $\dfrac{\partial z}{\partial \theta}$.

Using the chain rule

$$\frac{\partial z}{\partial r} = \frac{\partial z}{\partial x} \cdot \frac{\partial x}{\partial r} + \frac{\partial z}{\partial y} \cdot \frac{\partial y}{\partial r}$$

$$= \frac{\partial z}{\partial x} \cos \theta + \frac{\partial z}{\partial y} \sin \theta$$

Similarly, differentiating with respect to θ

$$\frac{\partial z}{\partial \theta} = \frac{\partial z}{\partial x} \cdot \frac{\partial x}{\partial \theta} + \frac{\partial z}{\partial y} \cdot \frac{\partial y}{\partial \theta}$$

$$= -\frac{\partial z}{\partial x} r \sin \theta + \frac{\partial z}{\partial y} r \cos \theta$$

These two equations can now be solved to give

$$\frac{\partial z}{\partial x} = \frac{\partial z}{\partial r} \cos \theta - \frac{1}{r} \cdot \frac{\partial z}{\partial \theta} \sin \theta$$

and

$$\frac{\partial z}{\partial y} = \frac{\partial z}{\partial r} \sin \theta + \frac{1}{r} \cdot \frac{\partial z}{\partial \theta} \cos \theta$$

(ii) If $f(x, y) = x^4 + xy + y^2$ where $x = 2at$, $y = at^2$ find $\mathrm{d}f/\mathrm{d}t$. From the chain rule

$$\frac{\mathrm{d}f}{\mathrm{d}t} = \frac{\partial f}{\partial x} \cdot \frac{\mathrm{d}x}{\mathrm{d}t} + \frac{\partial f}{\partial y} \cdot \frac{\mathrm{d}y}{\mathrm{d}t}$$

$$= (4x^3 + y) \cdot \frac{\mathrm{d}x}{\mathrm{d}t} + (x + 2y) \cdot \frac{\mathrm{d}y}{\mathrm{d}t}$$

$$= (4 \cdot 8a^3 t^3 + at^2) \cdot 2a + (2at + 2at^2)2at$$

$$= 2a^2 t^2 [3 + 2(1 + 16a^2)t]$$

This result can be verified by direct substitution. Let $x = 2at$ and $y = at^2$ in $f(x, y) = x^4 + xy + y^2$, then

$$f = 16a^4 t^4 + 2a^2 t^3 + a^2 t^4$$

102

so that

$$df/dt = 64a^4t^3 + 6a^2t^2 + 4a^2t^3$$
$$= 2a^2t^2[3 + 2(1 + 16a^2)t]$$

4.8 Euler's Theorem for Homogeneous Functions

If $f(x, y)$ is a homogeneous differentiable function of degree N, that is a function such that $f(\lambda X, \lambda Y) = \lambda^N f(X, Y)$, then

$$x\frac{\partial f}{\partial x} + y\frac{\partial f}{\partial y} = Nf(x, y) \tag{4.8.1}$$

PROOF   Let $u = tx, v = ty$. Then, since f is homogeneous with degree N,

$$f(u, v) = f(tx, ty) = t^N f(x, y) \tag{4.8.2}$$

Differentiating with respect to t

$$\frac{\partial f}{\partial t} = Nt^{N-1}f(x, y) \tag{4.8.3}$$

Alternatively

$$\frac{\partial f}{\partial t} = \frac{\partial f}{\partial u}\cdot\frac{\partial u}{\partial t} + \frac{\partial f}{\partial v}\cdot\frac{\partial v}{\partial t}$$
$$= x\frac{\partial f}{\partial u} + y\frac{\partial f}{\partial v} \tag{4.8.4}$$

Hence Equations (4.8.3) and (4.8.4) give

$$x\frac{\partial f}{\partial u} + y\frac{\partial f}{\partial v} = Nt^{N-1}f(x, y) \tag{4.8.5}$$

Now, when $t = 1$, $\partial f/\partial u = \partial f/\partial x$ and $\partial f/\partial v = \partial f/\partial y$ therefore (4.8.5) with $t = 1$ gives

$$x\frac{\partial f}{\partial x} + y\frac{\partial f}{\partial y} = Nf(x, y) \tag{4.8.6}$$

In general, if $f(x_1, x_2, x_3, \ldots, x_n)$ is a homogeneous function of degree N in the n variables $x_1, x_2, \ldots, x_n$ then

$$\sum_{r=1}^{n} x_r\frac{\partial f}{\partial x_r} = x_1\frac{\partial f}{\partial x_1} + x_2\frac{\partial f}{\partial x_2} + \cdots + x_n\frac{\partial f}{\partial x_n} = Nf \tag{4.8.7}$$

103

Example

(i) Verify Euler's Theorem for the function $f(x, y) = x^2 \tan^{-1}(y/x)$.
    By direct substitution

$$f(\lambda X, \lambda Y) = \lambda^2 X^2 \tan^{-1} \lambda Y/\lambda X$$
$$= \lambda^2 X^2 \tan^{-1}(Y/X)$$
$$= \lambda^2 f(X, Y)$$

Hence the function is homogeneous of degree two. Differentiating with respect to x and y in turn

$$\frac{\partial f}{\partial x} = \frac{\partial}{\partial x} [x^2 \tan^{-1}(y/x)] = 2x \tan^{-1}(y/x) + x^2 \cdot \frac{-y/x^2}{1 + y^2/x^2}$$

$$= 2x \tan^{-1}(y/x) - \frac{x^2 y}{x^2 + y^2}$$

and

$$\frac{\partial f}{\partial y} = \frac{\partial}{\partial y} [x^2 \tan^{-1}(y/x)] = x^2 \frac{1/x}{1 + y^2/x^2} = \frac{x^3}{x^2 + y^2}$$

Therefore it follows that

$$x \frac{\partial f}{\partial x} + y \frac{\partial f}{\partial y} = 2x^2 \tan^{-1}(y/x)$$

$$= 2f(x, y)$$

and Euler's theorem holds.

4.9 Implicit Functions

In many situations the relationship between three variables x, y and z cannot be expressed as an explicit function

$$z = f(x, y) \tag{4.9.1}$$

but is more conveniently, indeed in some cases can only be written implicitly as

$$F(x, y, z) = 0 \tag{4.9.2}$$

For this type of functional relationship the calculation of the partial derivatives is slightly more complicated. In all the previous work derivatives are obtained by a limiting process applied directly to a function $z = f(x, y)$, but here no such function can be written down and the differentiation process must be indirect.

There is an implicit relationship between x, y and z, therefore any incremental changes dx and dy will give rise to a change dz given by

$$dz = \frac{\partial z}{\partial x} dx + \frac{\partial z}{\partial y} dy \qquad (4.9.3)$$

The relationship between x, y and z is expressed as

$$F(x, y, z) = 0$$

and any changes in x, y, z must preserve this identity. Therefore any increment dx, dy, dz must produce a zero increment in F giving

$$0 = dF = \frac{\partial F}{\partial x} dx + \frac{\partial F}{\partial y} dy + \frac{\partial F}{\partial z} dz \qquad (4.9.4)$$

Now dz can be eliminated between (4.9.3) and (4.9.4) to give

$$\left(\frac{\partial F}{\partial x} + \frac{\partial F}{\partial z} \cdot \frac{\partial z}{\partial x} \right) dx + \left(\frac{\partial F}{\partial y} + \frac{\partial F}{\partial z} \cdot \frac{\partial z}{\partial y} \right) dy = 0 \qquad (4.9.5)$$

The variables x and y are independent, therefore the coefficients of dx and dy in this expression must vanish identically, that is

$$\frac{\partial F}{\partial x} + \frac{\partial F}{\partial z} \cdot \frac{\partial z}{\partial x} = 0 \qquad \text{and} \qquad \frac{\partial F}{\partial y} + \frac{\partial F}{\partial z} \cdot \frac{\partial z}{\partial y} = 0 \qquad (4.9.6)$$

Hence, the partial derivatives of z are given as

$$\frac{\partial z}{\partial x} = -\frac{\partial F}{\partial x} \bigg/ \frac{\partial F}{\partial z} \qquad \text{and} \qquad \frac{\partial z}{\partial y} = -\frac{\partial F}{\partial y} \bigg/ \frac{\partial F}{\partial z} \qquad (4.9.7)$$

Higher derivatives can be found by differentiating these expressions.

Example Given that $z^3 - z + x^2 + y^2 = 0$, find $\partial^2 z / \partial x \partial y$.
   The variables x, y, z are related by the implicit function

$$z^3 - z + x^2 + y^2 = 0$$

Differentiation of this expression with respect to x, gives

$$(3z^2 - 1) \frac{\partial z}{\partial x} + 2x = 0$$

so that

$$\frac{\partial z}{\partial x} = -\frac{2x}{3z^2 - 1}$$

Similarly, differentiation with respect to y, gives

$$\frac{\partial z}{\partial y} = -\frac{2y}{3z^2 - 1}$$

Hence

$$\frac{\partial^2 z}{\partial x \partial y} = \frac{\partial}{\partial x}\left(\frac{\partial z}{\partial y}\right)$$

$$= \frac{\partial}{\partial x}\left(-\frac{2y}{3z^2 - 1}\right)$$

$$= -2y\frac{\partial}{\partial x}\left(\frac{1}{3z^2 - 1}\right)$$

$$= 2y\frac{1}{(3z^2 - 1)^2}\frac{\partial}{\partial x}(3z^2 - 1)$$

$$= \frac{2y}{(3z^2 - 1)^2}6z\frac{\partial z}{\partial x}$$

$$= -\frac{12yz}{(3z^2 - 1)^2}\cdot\frac{2x}{(3z^2 - 1)}$$

That is

$$\frac{\partial^2 z}{\partial x \partial y} = -\frac{24xyz}{(3z^2 - 1)^3}$$

4.10 Jacobians

Let four variables x, y, u, v be connected by the implicit relations

$$F(x, y, u, v) = 0 \qquad G(x, y, u, v) = 0 \qquad (4.10.1)$$

Under certain conditions these two relationships imply that u and v can be regarded as functions of x and y and their partial derivatives can be evaluated.

106

Since F and G are identically zero it follows that

$$0 = \mathrm{d}F = \frac{\partial F}{\partial x}\,\mathrm{d}x + \frac{\partial F}{\partial y}\,\mathrm{d}y + \frac{\partial F}{\partial u}\,\mathrm{d}u + \frac{\partial F}{\partial v}\,\mathrm{d}v$$

and   (4.10.2)

$$0 = \mathrm{d}G = \frac{\partial G}{\partial x}\,\mathrm{d}x + \frac{\partial G}{\partial y}\,\mathrm{d}y + \frac{\partial G}{\partial u}\,\mathrm{d}u + \frac{\partial G}{\partial v}\,\mathrm{d}v$$

The functions F and G implicitly determine u and v as functions of x and y; therefore the increments are related by

$$\mathrm{d}u = \frac{\partial u}{\partial x}\,\mathrm{d}x + \frac{\partial u}{\partial y}\,\mathrm{d}y$$

                                                                                       (4.10.3)

$$\mathrm{d}v = \frac{\partial v}{\partial x}\,\mathrm{d}x + \frac{\partial v}{\partial y}\,\mathrm{d}y$$

The four identities (4.10.2) and (4.10.3) give

$$\left(\frac{\partial F}{\partial x} + \frac{\partial F}{\partial u}\cdot\frac{\partial u}{\partial x} + \frac{\partial F}{\partial v}\cdot\frac{\partial v}{\partial x}\right)\mathrm{d}x + \left(\frac{\partial F}{\partial y} + \frac{\partial F}{\partial u}\cdot\frac{\partial u}{\partial y} + \frac{\partial F}{\partial v}\cdot\frac{\partial v}{\partial y}\right)\mathrm{d}y = 0$$

and   (4.10.4)

$$\left(\frac{\partial G}{\partial x} + \frac{\partial G}{\partial u}\cdot\frac{\partial u}{\partial x} + \frac{\partial G}{\partial v}\cdot\frac{\partial v}{\partial x}\right)\mathrm{d}x + \left(\frac{\partial G}{\partial y} + \frac{\partial G}{\partial u}\cdot\frac{\partial u}{\partial y} + \frac{\partial G}{\partial v}\cdot\frac{\partial v}{\partial y}\right)\mathrm{d}y = 0$$

Now x and y are independent, therefore

$$\frac{\partial F}{\partial x} + \frac{\partial F}{\partial u}\cdot\frac{\partial u}{\partial x} + \frac{\partial F}{\partial v}\cdot\frac{\partial v}{\partial x} = 0 \qquad \frac{\partial G}{\partial x} + \frac{\partial G}{\partial u}\cdot\frac{\partial u}{\partial x} + \frac{\partial G}{\partial v}\cdot\frac{\partial v}{\partial x} = 0$$

                                                                                       (4.10.5)

and

$$\frac{\partial F}{\partial y} + \frac{\partial F}{\partial u}\cdot\frac{\partial u}{\partial y} + \frac{\partial F}{\partial v}\cdot\frac{\partial v}{\partial y} = 0 \qquad \frac{\partial G}{\partial y} + \frac{\partial G}{\partial u}\cdot\frac{\partial u}{\partial y} + \frac{\partial G}{\partial v}\cdot\frac{\partial v}{\partial y} = 0$$

                                                                                       (4.10.6)

Equations (4.10.5) are a pair of simultaneous equations that determine $\partial u/\partial x$ and $\partial v/\partial x$. Similarly Equations (4.10.6) determine $\partial u/\partial y$ and $\partial v/\partial y$.

These two pairs of equations have a unique solution provided that

$$J = \frac{\partial F}{\partial u}\cdot\frac{\partial G}{\partial v} - \frac{\partial F}{\partial v}\cdot\frac{\partial G}{\partial u} \neq 0$$ (4.10.7)

This expression is called the Jacobian of F, G with respect to u, v and is written as

$$J = \frac{\partial(F,G)}{\partial(u,v)} \tag{4.10.8}$$

Provided that the Jacobian is non-zero, the partial derivatives can be found from (4.10.5) (4.10.6) and written, using the Jacobian notation, as

$$\frac{\partial u}{\partial x} = -\frac{\partial(F,G)}{\partial(x,v)} \bigg/ \frac{\partial(F,G)}{\partial(u,v)} \qquad \frac{\partial v}{\partial x} = -\frac{\partial(F,G)}{\partial(u,x)} \bigg/ \frac{\partial(F,G)}{\partial(u,v)} \tag{4.10.9}$$

and

$$\frac{\partial u}{\partial y} = -\frac{\partial(F,G)}{\partial(y,v)} \bigg/ \frac{\partial(F,G)}{\partial(u,v)} \qquad \frac{\partial v}{\partial y} = -\frac{\partial(F,G)}{\partial(u,y)} \bigg/ \frac{\partial(F,G)}{\partial(u,v)} \tag{4.10.10}$$

4.11   Error Analysis

In any experiment where a certain quantity has to be determined in terms of given parameters there is a built in error owing to the limits of accuracy of the measurement of the parameters. It is therefore important that a method is available to estimate this type of error. The concept of small increments introduced in Section 4.6 becomes a powerful tool in this respect.

The principle is demonstrated in the following practical examples.

Examples

(i)   The volume of a circular cylinder of radius r and length L is given by the formula

$$V = \pi r^2 L \tag{4.11.1}$$

If r and L are measured to be 40 mm and 100 mm, respectively, but these measurements are subject to a possible error of 0.5 mm, find the percentage error in V.

The volume V is a function of the two variables r and L, that is

$$V = V(r, L) \tag{4.11.2}$$

Suppose that there are small errors in r and L denoted by Δr and ΔL, respectively. Using Formula (4.6.9) the increment ΔV

caused by these errors is

$$\Delta V \doteq \frac{\partial V}{\partial r} \Delta r + \frac{\partial V}{\partial L} \Delta L \tag{4.11.3}$$

On differentiation of the expression (4.11.1) it follows that

$$\frac{\partial V}{\partial r} = 2\pi rL \quad \text{and} \quad \frac{\partial V}{\partial L} = \pi r^2 \tag{4.11.4}$$

so that (4.11.3) becomes

$$\Delta V = 2\pi rL\, \Delta r + \pi r^2\, \Delta L \tag{4.11.5}$$

The percentage error in V is now easily calculated.

$$\text{Percentage error in } V = \frac{\Delta V}{V} \times 100$$

$$= \frac{1}{\pi r^2 L}(2\pi rL \cdot \Delta r + \pi r^2 \cdot \Delta L) \times 100$$

$$= \left(2\frac{\Delta r}{r} + \frac{\Delta L}{L}\right) \times 100$$

$$= 2 \times \text{percentage error in } r$$
$$+ \text{percentage error in } L \tag{4.11.6}$$

In this particular example

$$\text{Percentage error in } V = 2 \times \frac{0.5}{40} \times 100 + \frac{0.5}{100} \times 100$$

$$= 3\%$$

(ii) In a meter bridge experiment to find the resistance of a wire using a known resistance r the balance point occurs at a distance x from one end of the wire. The resistance R of the wire is then given by the formula

$$R = \frac{rx}{1000 - x} \tag{4.11.7}$$

If $r = 10$ ohms ($\pm 1\%$) and $x = 600$ mm (± 1 mm), find R and the limits of accuracy of the results.

Direct substitution of the given values into (4.11.7) gives

$$R = 15 \text{ ohms}$$

Let the errors in r and x be Δr and Δx, respectively. Then the error in R using (4.6.9) is,

$$\Delta R \doteq \frac{\partial R}{\partial r} \Delta r + \frac{\partial R}{\partial x} \Delta x \qquad (4.11.8)$$

Differentiating (4.11.7)

$$\frac{\partial R}{\partial r} = \frac{x}{1000 - x} \quad \text{and} \quad \frac{\partial R}{\partial x} = \frac{2r(500 - x)}{(1000 - x)^2} \qquad (4.11.9)$$

then (4.11.8) becomes

$$\Delta R = \frac{x}{(1000 - x)} \Delta r + \frac{2r(500 - x)}{(1000 - x)^2} \Delta x \qquad (4.11.10)$$

For the given example $r = 10$ ohms, $x = 600$ mm, $\Delta r = \pm 0.01 \times r = \pm 0.1$ ohms and $\Delta x = \pm 1$ mm.

With this data the largest value of ΔR is

$$\Delta R = \frac{600}{400} \times \frac{1}{10} + \frac{2 \times 10 \times (-100) \times (-1)}{400^2}$$

$$= \frac{3}{20} + \frac{1}{80} = \frac{13}{80}$$

Therefore the largest percentage error is

$$\frac{\Delta R}{R} \times 100 = \frac{13}{80} \times \frac{100}{15}$$

$$= 1.0833\%$$

Hence the result is accurate to within limits $\pm 1.0833\%$.

In the two worked examples the functions involved are functions of two variables. The method is easily extended when the quantity under consideration depends on many parameters by application of the incremental formula

$$\Delta f(x_1, x_2, \ldots, x_n) = \sum_{i=1}^{n} \frac{\partial f}{\partial x_i} \Delta x_i \qquad (4.11.11)$$

Exercises

4.1   Find $\lim\limits_{\substack{x \to 0 \\ y \to 0}} \left(\dfrac{\sin x + \sin y}{x + y} \right)$

4.2   Discuss the behaviour of the function

$$f(x, y) = \frac{x - y}{x + y} \qquad (x, y) \neq (0, 0)$$

$$f(0, 0) = 0$$

as $(x, y) \to (0, 0)$.

4.3   Find $\partial f/\partial x$ and $\partial f/\partial y$ for the functions
 (i) $f(x, y) = 4x^2 + 3y^2 - x + y$
 (ii) $f(x, y) = \exp(\sin x + \sin y)$
 (iii) $f(x, y) = \tan xy$
 (iv) $f(x, y) = \sin^{-1}(y/x)$
 (v) $f(x, y) = \ln \sin xy$

4.4   If $z = e^{xy}$ find $\partial^2 z/\partial x \partial y$.

4.5   Find $f_x(1, 2)$ and $f_y(-1, -2)$ for the function $f(x, y) = (x + y)(x - y)^2$.

4.6   Given that $f(x, y) = x^2 \sin y + y^2 \sin x$, find $\partial^2 f/\partial x^2$, $\partial^2 f/\partial x \partial y$, $\partial^2 f/\partial y^2$.

4.7   If $V = (x^2 + y^2 + z^2)^{-1/2}$ show that

$$\frac{\partial^2 V}{\partial x^2} + \frac{\partial^2 V}{\partial y^2} + \frac{\partial^2 V}{\partial z^2} = 0$$

4.8   If $u = e^x \cosh y, v = e^x \sinh y$ prove that

$$\frac{\partial^2 f}{\partial u^2} - \frac{\partial^2 f}{\partial v^2} = e^{-2x}\left(\frac{\partial^2 f}{\partial x^2} - \frac{\partial^2 f}{\partial y^2}\right)$$

4.9   Given that $U = f(t + r)/r + g(t - r)/r$ where $r^2 = x^2 + y^2 + z^2$ and f, g are functions of a single variable, show that

$$\frac{\partial^2 U}{\partial x^2} + \frac{\partial^2 U}{\partial y^2} + \frac{\partial^2 U}{\partial z^2} = \frac{\partial^2 U}{\partial t^2}$$

4.10 Given that $x = u^2 - v^2$ and $y = 2uv$ find

 (i) $\left(\dfrac{\partial x}{\partial u}\right)_v$ (ii) $\left(\dfrac{\partial y}{\partial v}\right)_u$ (iii) $\left(\dfrac{\partial u}{\partial x}\right)_y$ (iv) $\left(\dfrac{\partial y}{\partial v}\right)_x$

4.11 If $f(x, y) = (x^2 + y^2)^{1/2} + xe^{y/x}$ show that

$$x\frac{\partial f}{\partial x} + y\frac{\partial f}{\partial y} = f$$

4.12 If $f(x, y) = F(r, \theta)$ where $x = r\cos\theta$ and $y = r\sin\theta$, show that

$$\frac{\partial^2 f}{\partial x^2} + \frac{\partial^2 f}{\partial y^2} = \frac{\partial^2 F}{\partial r^2} + \frac{1}{r}\frac{\partial F}{\partial r} + \frac{1}{r^2}\frac{\partial^2 F}{\partial \theta^2}$$

4.13 If $U = \frac{3}{2}RT - a/v$ and $pv + a/v = RT$ where a and R are constants, express dU in terms of dp and dT and hence or otherwise find

$$\left(\frac{\partial U}{\partial p}\right)_T \quad \text{and} \quad \left(\frac{\partial U}{\partial T}\right)_p$$

4.14 Show that if $x = \ln(uv)$, $y = \ln(u/v)$ and $f(x, y) = F(u, v)$ then

$$u\frac{\partial F}{\partial u} + v\frac{\partial F}{\partial v} = \frac{\partial f}{\partial x} \qquad u\frac{\partial F}{\partial u} - v\frac{\partial F}{\partial v} = \frac{\partial f}{\partial y}$$

4.15 If $u = e^x(x\cos y - y\sin y)$ and $v = e^x(x\sin y + y\cos y)$ show that

$$\frac{\partial v}{\partial y} = \frac{\partial u}{\partial x} = e^{2x}\{(x + 1)^2 + y^2\}\frac{\partial x}{\partial u}$$

Find

$$\frac{\partial(x, y)}{\partial(u, v)} \quad \text{and} \quad \frac{\partial(u, v)}{\partial(x, y)}$$

4.16 If $f(x, y)$ and $g(x, y)$ are two functions of x, y and $x = x(u, v)$ and $y = y(u, v)$ show that

$$\frac{\partial(f, g)}{\partial(x, y)} = \frac{\partial(f, g)}{\partial(u, v)} \cdot \frac{\partial(u, v)}{\partial(x, y)}$$

Hence show that

$$\frac{\partial(u, v)}{\partial(x, y)} = 1 \bigg/ \frac{\partial(x, y)}{\partial(u, v)}$$

4.17 If $f(x, y) = (x - 2y)(x + y)^2$ where $x = at^2$, $y = 2at$ show that

$$\frac{df}{dt} = 6a^3 t(t + 2)(t^2 - 2t + 4)$$

4.18 When a projectile is projected with speed V at an angle α to the horizontal, its horizontal range R and its time of flight T are given by

$$R = \frac{V^2}{g} \sin 2\alpha \qquad T = \frac{2V}{g} \sin \alpha$$

where g is a constant.

The quantities V and g can be measured accurately but the angle can only be measured to within an error $\Delta\alpha$, find the respective errors in R and T.

If $\alpha = 30°$ show that the percentage error in T is $1\frac{1}{2}$ times that in R.

4.19 The lengths of the two adjacent sides of a right-angled triangle are measured to be 5 mm and 12 mm with a possible error of 1% find the percentage error in the calculated length of the hypotenuse.

4.20 A vertical pole mounted on a tower is observed to subtend equal angles at two points on level ground distant L apart in line with the foot of the tower on the same side of it. The angles of elevation of the lowest point of the pole from the two points are α and $\beta(\alpha > \beta)$ giving the height h of the tower as

$$h = \frac{L \cos(\alpha + \beta)}{\sin(\alpha - \beta)}$$

If α and β are measured to be 55° and 25°, respectively, accurate to within 1%, find the percentage error in h.

5
Taylor Series

5.1 Introduction

The most elementary functions in analysis are the rational functions formed by a finite number of applications of the basic laws of algebra, namely, addition, subtraction, multiplication and division. In particular the simplest of all are the polynomials, being a finite linear combination of integral powers of a variable x, that is functions of the form

$$f(x) = c_0 + c_1 x + c_2 x^2 + \cdots + c_n x^n \tag{5.1.1}$$

where $c_0, c_1, \ldots, c_n$ are constants.

The question of how or whether general functions can be approximated by polynomials is important both in the theory and in practical applications.

A simple approximation is given by Taylor's theorem in the following section, initially for functions of single variable and in later stages for functions of more variables.

5.2 Taylor's Theorem for Functions of One Variable

Suppose in the first instance that $f(x)$ is a polynomial of degree n. In this case $f(x)$ is of the form

$$f(x) = c_0 + c_1 x + c_2 x^2 + c_3 x^3 + \cdots + c_n x^n \tag{5.2.1}$$

and it is an easy matter to determine the coefficients.

By repeated differentiation of both sides of the identity (5.2.1) and setting $x = 0$ it is seen that

$$c_0 = f(0) \qquad c_1 = f'(0) \qquad c_2 = \frac{f''(0)}{2!}, \ldots, \qquad c_n = \frac{f^{(n)}(0)}{n!}$$

$$\tag{5.2.2}$$

Hence

$$f(x) = f(0) + f'(0)x + \frac{f''(0)x^2}{2!} + \cdots + \frac{f^{(n)}(0)x^n}{n!} \qquad (5.2.3)$$

This is Taylor's theorem for polynomials, and the result does no more than give a particular interpretation of the coefficients.

The theorem can be generalised by replacing x by $x + \alpha$ where α is a constant. Let

$$g(x) = f(x + \alpha) \qquad (5.2.4)$$

then since $f(x)$ is a polynomial of degree n, $g(x)$ is also a polynomial of degree n, so that

$$g(x) = g(0) + g'(0)x + \frac{g''(0)x^2}{2!} + \cdots + \frac{g^{(n)}(0)x^n}{n!} \qquad (5.2.5)$$

Repeated differentiation of the relationship $g(x) = f(x + \alpha)$ gives

$$g'(x) = f'(x + \alpha) \qquad g''(x) = f''(x + \alpha),\ldots,$$
$$g^{(n)}(x) = f^{(n)}(x + \alpha) \qquad (5.2.6)$$

and consequently, setting $x = 0$

$$g'(0) = f'(\alpha) \qquad g''(0) = f''(\alpha),\ldots, \qquad g^{(n)}(0) = f^{(n)}(\alpha) \qquad (5.2.7)$$

Hence the expression (5.2.5) may be rewritten as

$$f(x + \alpha) = f(\alpha) + f'(\alpha)x + \frac{f''(\alpha)x^2}{2!} + \cdots + \frac{f^{(n)}(\alpha)x^n}{n!} \qquad (5.2.8)$$

or alternatively, writing $X = x + \alpha$

$$f(X) = f(\alpha) + f'(\alpha)(X - \alpha) + \frac{f''(\alpha)}{2!}(X - \alpha)^2 + \cdots$$
$$+ \frac{f^{(n)}(\alpha)}{n!}(X - \alpha)^n \qquad (5.2.9)$$

In the case of polynomials, the Taylor theorem gives an exact representation of the function.

If $f(x)$ is a general function the equivalent polynomial form can be used as an approximation to the function.

Let $f(x)$ be a function with continuous derivatives up to order n then the approximation can be expressed as

$$f(x) = f(\alpha) + f'(\alpha)(x - \alpha) + \frac{f''(\alpha)}{2!}(x - \alpha)^2 + \cdots$$
$$+ \frac{f^{(n)}(\alpha)}{n!}(x - \alpha)^n + R_n \qquad (5.2.10)$$

115

Here, R_n is the remainder and represents the difference between the function $f(x)$ and its polynomial approximation. For a given value of x the remainder R_n is a function of α, denoted by $R_n = R_n(\alpha)$.

In the identity (5.2.10), let x be fixed. Then differentiation of the identity with respect to α gives

$$0 = \frac{f^{(n+1)}(\alpha)}{n!} (x - \alpha)^n + R'_n(\alpha) \qquad (5.2.11)$$

the rest of the terms cancelling out. Hence

$$R'_n(\alpha) = -\frac{f^{(n+1)}(\alpha)}{n!} (x - \alpha)^n \qquad (5.2.12)$$

Direct integration of this result with respect to α, gives

$$R_n(\alpha) - R_n(x) = -\int_x^\alpha \frac{f^{(n+1)}(u)}{n!} (x - u)^n \, du \qquad (5.2.13)$$

From the definition (5.2.10) it is apparent that the remainder R_n vanishes when $\alpha = x$, so that

$$R_n(x) = 0 \qquad (5.2.14)$$

Therefore (5.2.13) gives

$$R_n(\alpha) = -\int_x^\alpha \frac{f^{(n+1)}(u)}{n!} (x - u)^n \, du \qquad (5.2.15)$$

In particular, if $\alpha = 0$

$$R_n(0) = -\int_0^x \frac{f^{(n+1)}(u)}{n!} (x - u)^n \, du \qquad (5.2.16)$$

The remainder is an essential part of the approximate formula for $f(x)$, since it gives a measure of the approximation. In view of this importance it is useful to have an estimate of the size of R_n. The simplest estimate is obtained from the mean value theorem.

The simple mean value theorem (3.4.1) gives

$$R_n(\alpha) = \frac{f^{(n+1)}(\xi)}{n!} (x - \xi)^n (x - \alpha) \qquad \text{where } \alpha < \xi < x \quad (5.2.17)$$

This formula was discovered by Cauchy and is known as Cauchy's form of the remainder.

A simpler form of the remainder, originally obtained by Lagrange, and consequently known as Lagrange's form, follows from the generalized mean value theorem.

116

If $F(u)$ and $G(u)$ are two functions of u, then the generalized mean value Theorem (3.4.8) gives

$$\int_\alpha^x F(u)G(u)\,du = F(\xi)\int_\alpha^x G(u)\,du, \qquad \text{where } \alpha < \xi < x$$

In particular if $F(u) = f^{(n+1)}(u)$ and $G(u) = (x - u)^n$, then application of this theorem in the integral (5.2.15) gives

$$\begin{aligned}
R_n &= \frac{f^{(n+1)}(\xi)}{n!}\int_\alpha^x (x - u)^n\,du \\
&= \frac{f^{(n+1)}(\xi)}{n!}\left[-\frac{(x - u)^{n+1}}{n + 1} \right]_\alpha^x \\
&= \frac{f^{(n+1)}(\xi)}{(n + 1)!}(x - \alpha)^{n+1} \qquad \text{where } \alpha < \xi < x \qquad (5.2.18)
\end{aligned}$$

Using the Lagrange form of the remainder, Taylor's theorem can now be written as

$$f(x) = f(\alpha) + f'(\alpha)(x - \alpha) + \frac{f''(\alpha)}{2!}(x - \alpha)^2 + \cdots$$

$$+ \frac{f^{(n)}(\alpha)}{n!}(x - \alpha)^n + \frac{f^{(n+1)}(\xi)}{(n + 1)!}(x - \alpha)^{n+1}$$

$$\text{where } \alpha < \xi < x \qquad (5.2.19)$$

If all the derivatives of $f(x)$ are continuous and if $R_n \to 0$ as $n \to \infty$ then $f(x)$ can be expressed as an infinite series

$$f(x) = \sum_{n=0}^\infty \frac{f^{(n)}(\alpha)}{n!}(x - \alpha)^n \qquad (5.2.20)$$

In this case the series is called the Taylor expansion of $f(x)$ or the Taylor series for $f(x)$.

The expansion when $\alpha = 0$, namely

$$f(x) = \sum_{n=0}^\infty \frac{f^{(n)}(0)}{n!}x^n \qquad (5.2.21)$$

is usually referred to as a Maclaurin series.

The convergence of these series is dependent on the behaviour of R_n as n increases, and the series may only be valid for certain values of x. The range of values of x, for which the expansion is valid, is called the radius of convergence.

117

The series is commonly referred to as the expansion for $f(x)$ 'about the point $x = \alpha$'.

5.3 Expansions of the Elementary Functions

The Exponential Function

One of the simplest examples of a Taylor expansion is obtained for the exponential function $f(x) = e^x$. In this case all the derivatives are equal to the function itself, therefore

$$f^{(n)}(0) = f(0) = 1 \tag{5.3.1}$$

The polynomial approximation to the function is therefore

$$e^x = 1 + x + \frac{x^2}{2!} + \cdots + \frac{x^n}{n!} + R_n \tag{5.3.2}$$

where

$$R_n = \frac{e^\xi}{(n+1)!} x^{n+1} \qquad 0 < |\xi| < |x| \tag{5.3.3}$$

In order to complete the expansion in the form of an infinite series it is necessary to examine the behaviour of R_n as n increases.

For a given value of x let N be an integer greater than $2|x|$, then for all values of $n > N$

$$\frac{|x|}{n} < \frac{|x|}{N} < \frac{1}{2} \tag{5.3.4}$$

Consequently

$$\frac{|x|^{n+1}}{(n+1)!} = \frac{|x|^N}{N!} \cdot \frac{|x|}{(N+1)} \frac{|x|}{(N+2)} \cdots \frac{|x|}{(n+1)}$$

$$< \frac{|x|^N}{N!} \cdot \frac{1}{2} \cdot \frac{1}{2} \cdots \frac{1}{2}$$

$$= \frac{|x|^N}{N!} \left(\frac{1}{2}\right)^{n+1-N}$$

That is

$$\frac{|x|^{n+1}}{(n+1)!} < \frac{|2x|^N}{N!} \left(\frac{1}{2}\right)^{n+1} \tag{5.3.5}$$

118

Now, since $0 < |\xi| < |x|$ it follows that

$$|e^\xi| < e^{|\xi|} < e^{|x|} \tag{5.3.6}$$

and therefore using (5.3.4), (5.3.5) and (5.3.6)

$$|R_n| = \frac{|e^\xi|}{(n+1)!}|x|^{n+1}$$
$$< \frac{e^{|x|}|2x|^N}{N!} \cdot \left(\frac{1}{2}\right)^{n+1} \tag{5.3.7}$$

The first part of this expression is independent of n and as n increases $\left(\frac{1}{2}\right)^{n+1}$ tends to zero, therefore

$$\lim_{n \to \infty} R_n = 0 \tag{5.3.8}$$

Hence the expansion for e^x is

$$e^x = 1 + x + \frac{x^2}{2!} + \frac{x^3}{3!} + \cdots = \sum_{n=0}^{\infty} \frac{x^n}{n!} \tag{5.3.9}$$

and the series converges for all values of x.

In particular, putting $x = 1$

$$e = 1 + 1 + \frac{1}{2!} + \frac{1}{3!} + \frac{1}{4!} + \cdots = \sum_{n=0}^{\infty} \frac{1}{n!} \tag{5.3.10}$$

By taking ten terms in the series the reader will easily obtain the value $e \doteqdot 2.71828$, given in Appendix 1.2. More accurate estimates of the number e can be found by taking more terms in the series.

Sine and Cosine Series

Let $f(x) = \sin x$. Then the derivatives are

$$f'(x) = \cos x \qquad f''(x) = -\sin x$$
$$f'''(x) = -\cos x \qquad \text{etc.} \tag{5.3.11}$$

In general

$$f^{(2r)}(x) = (-1)^r \sin x$$

and

$$f^{(2r+1)}(x) = (-1)^r \cos x \qquad r = 0, 1, 2, 3, \ldots \tag{5.3.12}$$

Hence the derivatives evaluated at $x = 0$ are

$$f^{(2r)}(0) = 0 \qquad f^{(2r+1)}(0) = (-1)^r \tag{5.3.13}$$

so that the coefficients of all the even powers of x, in the expansion of $f(x)$, vanish and the series contains only odd powers of x.

The polynomial representation of $f(x) = \sin x$ is therefore

$$\sin x = x - \frac{x^3}{3!} + \frac{x^5}{5!} - \cdots + \frac{(-1)^n}{(2n+1)!} x^{2n+1} + R_n \qquad (5.3.14)$$

where

$$R_n = \frac{(-1)^{n+1}}{(2n+3)!} x^{2n+3} \cos \xi \qquad 0 < |\xi| < |x| \qquad (5.3.15)$$

The behaviour of R_n for large values of n can be obtained by a method similar to that used in the previous section.

From (5.3.15)

$$
\begin{aligned}
|R_n| &= \frac{|x|^{2n+3}}{(2n+3)!} |\cos \xi| \\
&< \frac{|x|^{2n+3}}{(2n+3)!}
\end{aligned}
\qquad (5.3.16)
$$

For a given value of x let N be an integer greater than $2|x|$, then for all $n > N$

$$\frac{|x|}{n} < \frac{|x|}{N} < \frac{1}{2} \qquad (5.3.17)$$

and

$$
\begin{aligned}
\frac{|x|^{2n+3}}{(2n+3)!} &= \frac{|x|^N}{N!} \frac{|x|}{N+1} \frac{|x|}{N+2} \cdots \frac{|x|}{(2n+3)} \\
&< \frac{|x|^N}{N!} \left(\frac{1}{2}\right)^{2n+3-N} \\
&= \frac{|2x|^N}{N!} \left(\frac{1}{2}\right)^{2n+3}
\end{aligned}
\qquad (5.3.18)
$$

Hence, since $(|2x|^N/N!)$ is fixed and $\left(\frac{1}{2}\right)^{2n+3}$ tends to zero as n increases

$$\lim_{n \to \infty} R_n = 0 \qquad (5.3.19)$$

Thus the expansion is

$$\sin x = x - \frac{x^3}{3!} + \frac{x^5}{5!} - \frac{x^7}{7!} + \cdots = \sum_{n=0}^{\infty} \frac{(-1)^n x^{2n+1}}{(2n+1)!} \qquad (5.3.20)$$

and the series converges for all values of x.

Similarly it can be shown that

$$\cos x = 1 - \frac{x^2}{2!} + \frac{x^4}{4!} - \frac{x^6}{6!} + \cdots = \sum_{n=0}^{\infty} \frac{(-1)^n x^{2n}}{(2n)!} \qquad (5.3.21)$$

for all values of x.

The Binomial Expansion

One of the most important expansions in series is known as the binomial expansion, for the function

$$f(x) = (1 + x)^p \qquad (5.3.22)$$

For this function

$$f'(x) = p(1 + x)^{p-1} \qquad f''(x) = p(p - 1)(1 + x)^{p-2}, \ldots$$
$$f^{(n)}(x) = p(p - 1)(p - 2)\ldots(p - n + 1)(1 + x)^{p-n} \qquad (5.3.23)$$

The form of the expansion depends on the index p.

Positive Integral Powers Let $p = N$ where N is a positive integer, then

$$f^{(n)}(x) \equiv 0 \text{ for } n \geqslant N + 1 \qquad (5.3.24)$$

For $n \leqslant N$

$$f^{(n)}(x) = N(N - 1)(N - 2)\ldots(N - n + 1)(1 + x)^{N-n}$$

$$= \frac{N!}{(N - n)!}(1 + x)^{N-n} \qquad (5.3.25)$$

giving

$$f^{(n)}(0) = \frac{N!}{(N - n)!} \qquad (5.3.26)$$

The expansion is therefore

$$f(x) = (1 + x)^N = 1 + Nx + \frac{N!}{(N - 2)!2!} x^2 + \cdots$$

$$+ \frac{N!}{(N - n)!n!} x^n + \cdots + x^N \qquad (5.3.27)$$

that is for an integer power N

$$(1 + x)^N = \sum_{n=0}^{N} \frac{N!}{(N - n)!n!} x^n \qquad (5.3.28)$$

121

Powers Other than Positive Integers   If p is not a positive integer

$$f^{(n)}(0) = p(p-1)(p-2)\ldots(p-n+1) \qquad (5.3.29)$$

and the polynomial approximation is

$$(1+x)^p = 1 + px + \frac{p(p-1)}{2!}x^2 + \frac{p(p-1)(p-2)}{3!}x^3 + \cdots$$

$$+ \frac{p(p-1)\ldots(p-n+1)}{n!}x^n + R_n \qquad (5.3.30)$$

where

$$R_n = \frac{p(p-1)(p-2)\ldots(p-n)}{(n+1)!}(1+\xi)^{p-n-1}x^{n+1} \qquad 0 < \xi < x$$

$$(5.3.31)$$

The behaviour of R_n is a little more complicated than the previous examples, but it is not difficult to show that, provided that $|x| < 1$, $R_n \to 0$ as $n \to \infty$.

The binomial series expansion can be written as

$$(1+x)^p = 1 + px + \frac{p(p-1)}{2!}x^2 + \cdots = \sum_{n=0}^{\infty} \binom{p}{n}x^n \qquad (5.3.32)$$

where the binomial coefficients $\binom{p}{n}$ are defined by

$$\binom{p}{n} = \frac{p(p-1)\ldots(p-n+1)}{n!} \qquad \text{for } n > 0$$

and

$$\binom{p}{0} = 1 \qquad (5.3.33)$$

5.4   Combination of Taylor Series

The rules of arithmetic can be extended to Taylor series to give an important practical method for finding the series for complicated functions.

Suppose that $f(x)$ and $g(x)$ have Taylor expansions

$$f(x) = a_0 + a_1 x + a_2 x^2 + \cdots = \sum_{n=0}^{\infty} a_n x^n \qquad (5.4.1)$$

$$g(x) = b_0 + b_1 x + b_2 x^2 + \cdots = \sum_{n=0}^{\infty} b_n x^n \qquad (5.4.2)$$

convergent for a given range of values of x.

The arithmetic rules then give

$$f(x) \pm g(x) = c_0 + c_1 x + c_2 x^2 + \cdots = \sum_{n=0}^{\infty} c_n x^n \qquad (5.4.3)$$

where

$$c_n = a_n \pm b_n \qquad (5.4.4)$$

and

$$f(x)g(x) = d_0 + d_1 x + d_2 x^2 + \cdots = \sum_{n=0}^{\infty} d_n x^n \qquad (5.4.5)$$

where

$$d_n = a_0 b_n + a_1 b_{n-1} + a_2 b_{n-2} + \cdots + a_n b_0 = \sum_{r=0}^{n} a_r b_{n-r} \qquad (5.4.6)$$

The combined series are convergent in the range of values of x which is included in both the ranges of convergence of the original series for $f(x)$ and $g(x)$.

Example Find the Taylor series expansion for the function $e^x \sin x$.

Considered separately the Taylor expansions of e^x and $\sin x$ are given in Section 5.3 as

$$e^x = 1 + x + \frac{x^2}{2!} + \frac{x^3}{3!} + \frac{x^4}{4!} + \cdots \qquad (5.4.7)$$

and

$$\sin x = x - \frac{x^3}{3!} + \frac{x^5}{5!} - \frac{x^7}{7!} + \cdots \qquad (5.4.8)$$

Therefore combining (5.4.7) and (5.4.8) according to the rules (5.4.5) and (5.4.6) it follows that

$$e^x \sin x = \left(1 + x + \frac{x^2}{2!} + \frac{x^3}{3!} + \cdots \right)$$
$$\times \left(x - \frac{x^3}{3!} + \frac{x^5}{5!} - \frac{x^7}{7!} + \cdots \right)$$

123

$$e^x \sin x = 1 \cdot x + 1 \cdot 1 \cdot x^2 + \left[1 \left(-\frac{1}{3!} \right) + \frac{1}{2!} \cdot 1 \right] x^3$$

$$+ \left[1 \left(-\frac{1}{3!} \right) + \frac{1}{3!} \cdot 1 \right] x^4$$

$$+ \left[1 \cdot \frac{1}{5!} + \frac{1}{2!} \left(-\frac{1}{3!} \right) + \frac{1}{4!} \cdot 1 \right] x^5 + \cdots$$

$$= x + x^2 + \frac{1}{3} x^3 - \frac{1}{30} x^5 + \cdots \qquad (5.4.9)$$

Another important technique is that of substitution, by which the Taylor series for a given function is obtained from a simpler series which is known or more easily obtained. This technique is illustrated by a simple example. Consider the function $F(x) = 1/(1 + x^2)$. The function is well defined and can be differentiated everywhere so that the Taylor series could be obtained directly. However the reader will soon verify that repeated differentiation of this function becomes complicated and the determination of the coefficients in the series is not easy.

The Taylor series is most easily found by considering the function $f(x) = 1/(1 + x)$.

Repeated differentiation of this function gives

$$f(x) = \frac{1}{(1 + x)} \qquad f'(x) = \frac{-1}{(1 + x)^2} \qquad f''(x) = \frac{2}{(1 + x)^3}, \ldots$$

$$f^{(n)}(x) = \frac{(-1)^n n!}{(1 + x)^{n+1}} \qquad (5.4.10)$$

Hence

$$f(0) = 1 \qquad f'(0) = -1 \qquad f''(0) = 2, \ldots \qquad f^{(n)}(0) = (-1)^n n!$$
$$(5.4.11)$$

giving the Taylor series

$$\frac{1}{(1 + x)} = 1 - x + x^2 - x^3 + x^4 - \cdots = \sum_{n=0}^{\infty} (-1)^n x^n \qquad (5.4.12)$$

The series for $F(x) = 1/(1 + x^2)$ now follows, on replacing x by x^2 in (5.4.12), as

$$F(x) = 1/(1 + x^2) = 1 - x^2 + x^4 - x^6 + x^8 \cdots$$

$$= \sum_{n=0}^{\infty} (-1)^n x^{2n} \qquad (5.4.13)$$

124

5.5 Taylor's Theorem for a Function of Two Variables

The Taylor expansion obtained for a function of one variable can be simply extended to include functions of two, and indeed functions of many variables.

Let $f(x, y)$ be a function of two variables with continuous partial derivatives, then the Taylor expansion follows by direct application of the results already proved for one variable.

To find the expansion for $f(x, y)$ about the point (α, β) put $x = \alpha + ht$, $y = \beta + kt$ then

$$f(x, y) = f(\alpha + ht, \beta + kt) = F(t) \qquad (5.5.1)$$

where $F(t)$ is a function of the single variable t.

The Taylor expansion for the function $F(t)$ is

$$F(t) = F(0) + F'(0)t + \frac{F''(0)}{2!} t^2 + \cdots + \frac{F^{(n)}(0)}{n!} t^n + R_n \qquad (5.5.2)$$

where

$$R_n = \frac{F^{(n+1)}(\xi)}{(n+1)!} t^{n+1} \qquad 0 < \xi < t \qquad (5.5.3)$$

The chain rule can now be applied to express the derivatives of F in terms of the partial derivatives of $f(x, y)$. From (4.7.7)

$$F'(t) = \frac{dx}{dt} \cdot \frac{\partial f}{\partial x} + \frac{dy}{dt} \cdot \frac{\partial f}{\partial y} \qquad (5.5.4)$$

$$= h \frac{\partial f}{\partial x} + k \frac{\partial f}{\partial y} \qquad (5.5.5)$$

Hence

$$F'(0) = h \frac{\partial}{\partial x} f(\alpha, \beta) + k \frac{\partial}{\partial y} f(\alpha, \beta) \qquad (5.5.6)$$

Repeating this operation

$$F''(t) = \frac{d}{dt} F'(t) \qquad (5.5.7)$$

$$= \left(h \frac{\partial}{\partial x} + k \frac{\partial}{\partial y} \right) F'(t)$$

$$= h^2 \frac{\partial^2 f}{\partial x^2} + 2hk \frac{\partial^2 f}{\partial x \partial y} + k^2 \frac{\partial^2 f}{\partial y^2} \qquad (5.5.8)$$

125

thus

$$F''(0) = h^2 \frac{\partial^2}{\partial x^2} f(\alpha, \beta) + 2hk \frac{\partial^2}{\partial x \partial y} f(\alpha, \beta) + k^2 \frac{\partial^2}{\partial y^2} f(\alpha, \beta) \quad (5.5.9)$$

Hence it can be deduced that

$$F(t) = f(\alpha, \beta) + t \left\{ h \frac{\partial}{\partial x} f(\alpha, \beta) + k \frac{\partial}{\partial y} f(\alpha, \beta) \right\}$$

$$+ \frac{t^2}{2!} \left\{ h^2 \frac{\partial^2}{\partial x^2} f(\alpha, \beta) + 2hk \frac{\partial^2}{\partial x \partial y} f(\alpha, \beta) \right.$$

$$\left. + k^2 \frac{\partial^2}{\partial y^2} f(\alpha, \beta) \right\} + \cdots + \frac{t^n}{n!} \left(h \frac{\partial}{\partial x} + k \frac{\partial}{\partial y} \right)^n f(\alpha, \beta) + R_n$$

$$(5.5.10)$$

where

$$R_n = \frac{t^{n+1}}{(n+1)!} \left(h \frac{\partial}{\partial x} + k \frac{\partial}{\partial y} \right)^{n+1} f(\alpha + \xi h, \beta + \xi k) \qquad 0 < \xi < 1$$

$$(5.5.11)$$

Now $x = \alpha + ht$ and $y = \beta + kt$ so that $ht = x - \alpha$ and $kt = y - \beta$ and the Taylor series can be written as

$$f(x, y) = f(\alpha, \beta) + (x - \alpha) \frac{\partial}{\partial x} f(\alpha, \beta) + (y - \beta) \frac{\partial}{\partial y} f(\alpha, \beta)$$

$$+ \frac{1}{2!} \left\{ (x - \alpha)^2 \frac{\partial^2}{\partial x^2} f(\alpha, \beta) \right.$$

$$+ 2(x - \alpha)(y - \beta) \frac{\partial^2}{\partial x \partial y} f(\alpha, \beta)$$

$$\left. + (y - \beta)^2 \frac{\partial^2}{\partial y^2} f(\alpha, \beta) \right\} + \cdots \qquad (5.5.12)$$

The Taylor series for a function of many variables can be similarly obtained by application of the chain rule for a function of many variables.

Examples Find the Taylor expansion, omitting terms higher than second order, for the following functions about the point indicated

(i) $f(x, y) = \dfrac{1 + xy}{1 + x + y}$ \qquad about $(0, 0)$

126

(ii) $f(x, y) = \sin(x + 2y)$ about $(0, \pi/4)$

(i) $f(x, y) = \dfrac{1 + xy}{1 + x + y}$ $f(0,0) = 1$

$$\frac{\partial f}{\partial x} = \frac{(1 + x + y)y - (1 + xy)}{(1 + x + y)^2}$$

$$= \frac{y^2 + y - 1}{(1 + x + y)^2} \qquad \frac{\partial}{\partial x} f(0,0) = -1$$

$$\frac{\partial f}{\partial y} = \frac{x^2 + x - 1}{(1 + x + y)^2} \qquad \frac{\partial}{\partial y} f(0,0) = -1$$

$$\frac{\partial^2 f}{\partial x^2} = \frac{\partial}{\partial x}\left(\frac{\partial f}{\partial x}\right) = \frac{-2(y^2 + y - 1)}{(1 + x + y)^3} \qquad \frac{\partial^2}{\partial x^2} f(0,0) = 2$$

$$\frac{\partial^2 f}{\partial x \partial y} = \frac{3 + x + y + 2xy}{(1 + x + y)^3} \qquad \frac{\partial^2}{\partial x \partial y} f(0,0) = 3$$

$$\frac{\partial^2 f}{\partial y^2} = \frac{-2(x^2 + x - 1)}{(1 + x + y)^2} \qquad \frac{\partial^2}{\partial y^2} f(0,0) = 2$$

Hence the Taylor expansion up to and including second order terms is

$$f(x, y) = 1 - x - y + \frac{1}{2!}(2x^2 + 2 \cdot 3xy + 2y^2) + \cdots$$

$$= 1 - x - y + x^2 + 3xy + y^2 + \cdots$$

(ii) $f(x, y) = \sin(x + 2y)$ $f(0, \pi/4) = \sin\tfrac{1}{2}\pi = 1$

$$\frac{\partial f}{\partial x} = \cos(x + 2y) \qquad \frac{\partial}{\partial x} f(0, \pi/4) = \cos \pi/2 = 0$$

$$\frac{\partial f}{\partial y} = 2\cos(x + 2y) \qquad \frac{\partial}{\partial y} f(0, \pi/4) = 2\cos \pi/2 = 0$$

$$\frac{\partial^2 f}{\partial x^2} = -\sin(x + 2y) \qquad \frac{\partial^2}{\partial x^2} f(0, \pi/4) = -\sin \pi/2 = -1$$

$$\frac{\partial^2 f}{\partial x \partial y} = -2\sin(x + 2y) \qquad \frac{\partial^2}{\partial x \partial y} f(0, \pi/4) = -2\sin \pi/2 = -2$$

$$\frac{\partial^2 f}{\partial y^2} = -4\sin(x + 2y) \qquad \frac{\partial^2}{\partial y^2} f(0, \pi/4) = -4\sin \pi/2 = -4$$

Hence the Taylor expansion for $\sin(x + 2y)$ about the point $(0, \pi/4)$ is

$$f(x, y) = 1 + \frac{1}{2!}\left\{-x^2 + 2(-2)x(y - \pi/4)\right.$$

$$\left. - 4(y - \pi/4)^2\right\} + \cdots$$

that is

$$\sin(x + 2y) \doteq 1 - \frac{x^2}{2} - 2x(y - \pi/4) - 2(y - \pi/4)^2 + \cdots$$

5.6 Some Standard Series Expansions

$$e^x = 1 + x + \frac{x^2}{2!} + \cdots = \sum_{n=0}^{\infty} \frac{x^n}{n!} \qquad \text{all } x$$

$$\sin x = x - \frac{x^3}{3!} + \frac{x^5}{5!} - \frac{x^7}{7!} + \cdots = \sum_{n=0}^{\infty} (-1)^n \frac{x^{2n+1}}{(2n+1)!} \qquad \text{all } x$$

$$\cos x = 1 - \frac{x^2}{2!} + \frac{x^4}{4!} - \cdots = \sum_{n=0}^{\infty} \frac{(-1)^n x^{2n}}{(2n)!} \qquad \text{all } x$$

$$\ln(1 + x) = x - \frac{x^2}{2} + \frac{x^3}{3} - \frac{x^4}{4} + \cdots$$

$$= \sum_{n=1}^{\infty} \frac{(-1)^{n-1} x^n}{n} \qquad -1 < x \leqslant 1$$

$$\ln(1 - x) = -x - \frac{x^2}{2} - \frac{x^3}{3} - \frac{x^4}{4} - \cdots = -\sum_{n=1}^{\infty} \frac{x^n}{n} \qquad -1 \leqslant x < 1$$

$$\ln\left(\frac{1 + x}{1 - x}\right) = 2\left\{x + \frac{x^3}{3} + \frac{x^5}{5} + \cdots\right\} = 2 \sum_{n=0}^{\infty} \frac{x^{2n+1}}{2n + 1} \qquad -1 < x < 1$$

$$\frac{1}{1 - x} = 1 + x + x^2 + x^3 + \cdots = \sum_{n=0}^{\infty} x^n \qquad -1 < x < 1$$

$$\frac{1}{1 + x} = 1 - x + x^2 - x^3 + \cdots = \sum_{n=0}^{\infty} (-1)^n x^n \qquad -1 < x < 1$$

$$\frac{1}{\sqrt{(1 - x)}} = 1 + \tfrac{1}{2}x + \frac{1 \cdot 3}{2!} \frac{x^2}{4} + \frac{1 \cdot 3 \cdot 5}{3!} \frac{x^3}{8} + \cdots$$

$$= \sum_{n=0}^{\infty} \frac{(2n)!}{(n!)^2} \left(\frac{x}{4}\right)^n \qquad -1 \leqslant x < 1$$

Exercises

5.1 Find the first three terms of the Taylor expansions for each of the following functions in the neighbourhood of $x = 0$
   (i) $\cos^3 x$ (ii) $\sec x$ (iii) $\tan x$

5.2 Find the expansions for the functions
   (i) $\cosh x$ (ii) $\sinh x$ (iii) $\sin^2 x$
   (iv) $(1 - x^2)^{-1/2}$ (v) $\ln(1 + x^2)$ (vi) $\tan^{-1} x$
   (vii) $\sin^{-1} x$

5.3 Expand $(1 - x^2)^{-1/2}$ in powers of x up to the term in x^6. Hence by integration; or otherwise, expand $\cos^{-1} x$ in powers of x up to the term in x^7.

5.4 Given that

$$\frac{d}{dx} \sinh^{-1} x = \frac{1}{(1 + x^2)^{1/2}}$$

obtain the power series expansion

$$\sinh^{-1} x = x + \sum_{n=1}^{\infty} (-1)^n \frac{1 \cdot 3 \cdot 5 \ldots (2n - 1)}{2 \cdot 4 \cdot 6 \ldots 2n} \frac{x^{2n+1}}{2n + 1}$$

5.5 Find the Taylor expansion for $\tanh x$ in powers of x up to the term in x^3.

5.6 Expand

$$\frac{1 + x + 3x^2}{(1 + x)(1 + 2x^2)}$$

as a power series. Give the terms as far as that in x^4 and then obtain the general series.

5.7 Find the first three terms in the Taylor expansions for the following functions in the neighbourhood of $(0, 0)$

   (i) $\dfrac{1}{1 + \sin x + \sin y}$ (ii) $e^{(1 + x)(1 + y)}$

   (iii) $\ln\left(\dfrac{1 + x + y}{1 + x - y}\right)$

129

6
Maxima and Minima

6.1 Introduction

An important application of the differential calculus is in the determination of extreme values of a function of one or more variables. In this chapter the theory of maxima and minima is discussed in detail for functions of one and two variables. The procedure for more difficult problems with functions of more than two variables is discussed briefly.

6.2 Functions of One Variable

DEFINITIONS A continuous function $f(x)$ has a maximum at a point $x = \alpha$ if $f(x) < f(\alpha)$ for all values of $x \neq \alpha$ lying in some interval $\alpha - \delta < x < \alpha + \delta$.

A continuous function $f(x)$ has a minimum at a point $x = \alpha$ if $f(x) > f(\alpha)$ for all values of $x \neq \alpha$ lying in some interval $\alpha - \delta < x < \alpha + \delta$.

From Figure 6.1 it can be seen that the particular function represented by the curve $y = f(x)$ has maxima at the points P_1, P_3 and minima at the points P_2, P_4. Moreover the minimum value of $f(x)$ at P_4 is greater than the maximum value at P_1, thus the concept of maximum or minimum is relative and the points must always be regarded as local maxima or minima.

The greatest value of a continuous function $f(x)$ defined in a range $a \leqslant x \leqslant b$ must occur at either a local maximum or at one of the end points $x = a$ or $x = b$. Similarly the least value must occur at a local minimum or at one of the end points.

It has already been shown in the proof of Rolle's theorem in Chapter 2 that if a function has a local maximum or minimum at $x = \alpha$, then

$$f'(\alpha) = 0 \tag{6.2.1}$$

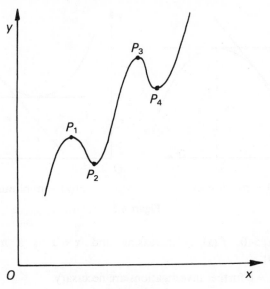

Figure 6.1

Hence, $f'(\alpha) = 0$ is a necessary condition for a continuous function $f(x)$ to have either a maximum or minimum at the point $x = \alpha$. It can be demonstrated by example that the condition is not a sufficient condition.

The function $f(x) = x^3$, satisfies $f'(0) = 0$ but $x = 0$ is neither a maximum nor a minimum point since $f(x) > 0$ for $x > 0$, and $f(x) < 0$ for $x < 0$.

Points at which $f'(x) = 0$ are called extreme points or stationary points or turning points. These points are easily located by solving the Equation $f'(x) = 0$, but further criteria are needed to determine their nature.

If $f(x)$ has a maximum at the point $x = \alpha$, then the curve $y = f(x)$ lies below the tangent at $x = \alpha$, as illustrated in Figure 6.2. In the neighbourhood of this maximum point, the slope of the tangent, or $f'(x)$, must decrease as the value of x increases. Similarly if $f(x)$ has a minimum at $x = \alpha$ then the curve lies above the tangent at $x = \alpha$ and the derived function $f'(x)$ is an increasing function in the neighbourhood of $x = \alpha$. Hence

if $f''(\alpha) < 0$, $f'(x)$ is decreasing and $x = \alpha$ is a maximum point;

(6.2.2)

131

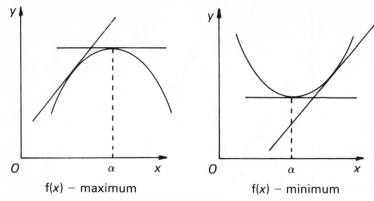

f(x) – maximum   f(x) – minimum

Figure 6.2

if $f''(\alpha) > 0$, $f'(x)$ is increasing and $x = \alpha$ is a minimum
point;   (6.2.3)

if $f''(\alpha) = 0$ further investigations are necessary.

For example, the three functions

$$f_1(x) = x^4 \qquad f_2(x) = 1 - x^4 \qquad f_3(x) = x^3 \qquad (6.2.4)$$

all satisfy the conditions

$$f'(0) = f''(0) = 0 \qquad (6.2.5)$$

but their behaviour at the point $x = 0$ is entirely different. At $x = 0$,

$f_1(x)$ has a minimum

$f_2(x)$ has a maximum

$f_3(x)$ has neither a maximum nor a minimum

The function $f_3(x) = x^3$, is said to have a point of inflexion at the
point $x = 0$.

DEFINITION   A point where $f''(x) = 0$ but the function has neither a
maximum nor a minimum is called a point of inflexion. At a point of
inflexion the first derivative does not necessarily vanish.

For example, the functions

$$f_1(x) = x^3 \qquad \text{and} \qquad f_2(x) = x^3 + x$$

both have points of inflexion at the origin $x = 0$ (Figure 6.3). In the
first case $f'_1(0) = 0$ but in the second case $f'_2(0) = 1$.

The important feature at a point of inflexion is that the gradient, or
derivative $f'(x)$, has either a maximum or minimum at that point.

132

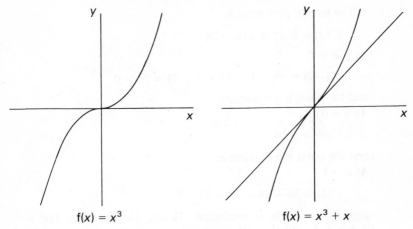

$$f(x) = x^3 \qquad\qquad\qquad f(x) = x^3 + x$$

Figure 6.3

The process of locating maximum or minimum points for a function $f(x)$ of a single variable can be summarized as follows.

(i) The points are found by solving the equation $f'(\alpha) = 0$.

(ii) The nature of the points is determined by examination of the second derivative $f''(x)$.

If $f''(\alpha) < 0$, then $x = \alpha$ is a maximum point.

If $f''(\alpha) > 0$, then $x = \alpha$ is a minimum point.

If $f''(\alpha) = 0$ the point may be either a maximum, a minimum or a point of inflexion and further investigation is required.

Examples

(i) Find the stationary points of the function

$$f(x) = 3x^4 + 4x^3 - 12x^2 + 1$$

and determine their nature.

For an extreme point the necessary condition is $f'(x) = 0$. In this case

$$f'(x) = 12x^3 + 12x^2 - 24x$$

so that the condition becomes

$$12x^3 + 12x^2 - 24x = 0$$

The equation can be factorized as

$$12x(x - 1)(x + 2) = 0$$

giving the stationary points $x = 0, x = 1, x = -2$.

133

The second derivative is

$$f''(x) = 36x^2 + 24x - 24$$

At $x = -2$

$$f''(-2) = 36 \times 4 - 24 \times 2 - 24 = 72 > 0$$

and the point is a minimum.
At $x = 0$

$$f''(0) = -24$$

and the point is a maximum.
At $x = 1$

$$f''(1) = 36 + 24 - 24 = 36 > 0$$

and the point is a minimum. Hence, the function $f(x) = 3x^4 + 4x^3 - 12x^2 + 1$ has a maximum at the point $x = 0$ and minima at the points $x = -2, x = 1$.

(ii) Find the stationary points of the function

$$f(x) = e^x(\cos x - \sin x)$$

Differentiating the given function

$$f'(x) = e^x(\cos x - \sin x - \sin x - \cos x)$$
$$= -2e^x \sin x$$

Therefore, $f'(x) = 0$ when $e^x \sin x = 0$. Now $e^x \neq 0$, therefore $\sin x = 0$, giving

$$x = n\pi \qquad n = 0, \pm 1, \pm 2, \pm 3 \qquad \text{etc.}$$

Differentiating a second time

$$f''(x) = -2e^x(\sin x + \cos x)$$

At the points $x = n\pi$

$$f''(n\pi) = -2e^{n\pi}(\sin n\pi + \cos n\pi)$$
$$= 2(-1)^{n+1} e^{n\pi}$$

If n is even

$$f''(n\pi) = -2e^{n\pi} < 0$$

If n is odd

$$f''(n\pi) = 2e^n > 0$$

Hence, the points

$$x = 0, \pm 2\pi, \pm 4\pi, \pm 6\pi, \ldots, \text{are maxima}$$

134

and the points

$$x = \pm\pi, \pm 3\pi, \pm 5\pi, \dots, \text{are minima}$$

6.3 Functions of Two Variables

DEFINITIONS   A continuous function $f(x, y)$ has a maximum at a point (α, β) if $f(x, y) < f(\alpha, \beta)$ for all points sufficiently close to (α, β).

A continuous function $f(x, y)$ has a minimum at a point (α, β) if $f(x, y) > f(\alpha, \beta)$ for all points sufficiently close to (α, β).

As in the case of functions of one variable the maxima and minima are relative and must be only regarded as local phenomena.

From these definitions it follows that if the function $f(x, y)$ has a maximum (minimum) at (α, β) then $f(x, \beta)$, treated as a function of one variable, has a maximum (minimum) at $x = \alpha$. Thus by the theory of functions of one variable a necessary condition for a function to have a maximum (minimum) at the point (α, β) is $f_x(\alpha, \beta) = 0$. Similarly it follows that $f_y(\alpha, \beta) = 0$. Although these conditions are necessary it is easily demonstrated that they are not sufficient.

The function $f(x, y) = x^2 - y^2$ represented in Figure 6.4 satisfies the condition $f_x(0, 0) = f_y(0, 0)$. In the plane $y = 0$, $f(x, y) = x^2$ and therefore has a minimum at $x = 0$, but in the plane $x = 0$, $f(x, y) = -y^2$ and the function has a maximum at $y = 0$. In this case the point is called a saddle point.

In general points where $f_x(x, y) = f_y(x, y) = 0$ are called stationary points.

In order to determine the nature of the stationary point at (α, β), it is necessary to compare the values of $f(\alpha, \beta)$ and $f(\alpha + h, \beta + k)$ for all values of h and k sufficiently small.

Let $f(x, y)$ be a function with continuous partial derivatives, then from Taylor's theorem, neglecting terms of order h^3, k^3 and higher,

$$f(\alpha + h, \beta + k) \doteq f(\alpha, \beta) + hf_x(\alpha, \beta) + kf_y(\alpha, \beta)$$
$$+ \tfrac{1}{2}\{h^2 f_{xx}(\alpha, \beta) + 2hk f_{xy}(\alpha, \beta) + k^2 f_{yy}(\alpha, \beta)\}$$

$$(6.3.1)$$

If (α, β) is a stationary point of $f(x, y)$, then $f_x(\alpha, \beta) = 0$ and $f_y(\alpha, \beta) = 0$ and thus

$$f(\alpha + h, \beta + k) - f(\alpha, \beta) = \tfrac{1}{2}\{h^2 f_{xx}(\alpha, \beta) + 2hk f_{xy}(\alpha, \beta) + k^2 f_{yy}(\alpha, \beta)\}$$

$$(6.3.2)$$

135

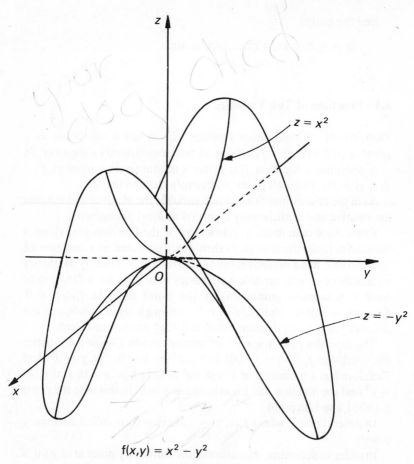

$$f(x,y) = x^2 - y^2$$

Figure 6.4

The nature of the stationary point depends on the sign of $f(\alpha + h, \beta + k) - f(\alpha, \beta)$ and therefore the sign of F where

$$F = h^2 f_{xx} + 2hk f_{xy} + k^2 f_{yy} \qquad (6.3.3)$$

Now, provided that $f_{xx} \neq 0$

$$F = \frac{1}{f_{xx}} \{ h^2 f_{xx}^2 + 2hk f_{xx} f_{xy} + k^2 f_{xx} f_{yy} \}$$

$$= \frac{1}{f_{xx}} \{ (hf_{xx} + kf_{xy})^2 + k^2 (f_{xx} f_{yy} - f_{xy}^2) \} \qquad (6.3.4)$$

136

The second term in (6.3.4) now becomes an important factor in the discussion. Let this combination of derivatives be written as

$$\Delta = f_{xx}f_{yy} - f_{xy}^2 \tag{6.3.5}$$

then there are three possibilities to consider, namely $\Delta > 0$, $\Delta < 0$ and $\Delta = 0$.

(i) $\Delta > 0$ If $f_{xx} > 0$, then it follows that $F > 0$. Hence, $f(\alpha + h, \beta + k) > f(\alpha, \beta)$ for all h, k so the point (α, β) is a minimum point. If $f_{xx} < 0$, then $F < 0$ giving $f(\alpha + h, \beta + k) < f(\alpha, \beta)$ for all h, k and (α, β) is a maximum point.

(ii) $\Delta < 0$ In this case, F can take either sign depending on the values of h and k. For example, if $k = 0$ then

$$F = h^2 f_{xx} \tag{6.3.6}$$

and F has the same sign as f_{xx}.
If $k = -hf_{xx}/f_{xy}$ then

$$F = \frac{k^2\Delta}{f_{xx}} = \frac{k^2\Delta}{f_{xx}^2} f_{xx} \tag{6.3.7}$$

Now $(k^2\Delta/f_{xx}^2) < 0$, therefore with this value of k, F has the opposite sign to f_{xx}.
Hence the stationary point is known as a saddle point.

(iii) $\Delta = 0$ In this event, the problem is best considered by methods other than Taylor series.
In all of the above discussion it has been assumed that $f_{xx}(\alpha, \beta) \neq 0$. If this condition is violated and $f_{xx}(\alpha, \beta) = 0$, then

$$F = k(2hf_{xy} + kf_{yy})$$

and F can be either positive or negative, depending on the values of h, k. This situation is covered by the case when $\Delta < 0$.

Summary

The stationary points of a continuous function $f(x, y)$ are located by the solution of the simultaneous equations

$$f_x(\alpha, \beta) = 0 \qquad f_y(\alpha, \beta) = 0$$

The nature of the stationary points is determined as follows. Compute, $f_{xx}(\alpha, \beta), f_{xy}(\alpha, \beta), f_{yy}(\alpha, \beta)$ and

$$\Delta = f_{xx}(\alpha, \beta)f_{yy}(\alpha, \beta) - f_{xy}^2(\alpha, \beta)$$

(i)   If $\Delta < 0$, the point is a saddle point
(ii) If $\Delta > 0$ and $f_{xx}(\alpha, \beta) < 0$ the point is a maximum point
(iii) If $\Delta > 0$ and $f_{xx}(\alpha, \beta) > 0$ the point is a minimum point
(iv) $\Delta = 0$ further investigation is required

Examples

(i)   Find the stationary points of

$$f(x, y) = x^4 - 4x^3 + 4x^2 - 3y^2 + 6y$$

and determine their nature.
      The first derivatives are

$$f_x(x, y) = 4x^3 - 12x^2 + 8x = 4x(x - 1)(x - 2) \qquad (6.3.8)$$

and

$$f_y(x, y) = -6y + 6 = -6(y - 1) \qquad (6.3.9)$$

At a stationary point $f_x(x, y) = f_y(x, y) = 0$, therefore

$$x(x - 1)(x - 2) = 0 \quad \text{and} \quad y - 1 = 0 \qquad (6.3.10)$$

giving $x = 0$, 1 or 2 and $y = 1$. $\qquad (6.3.11)$
Hence, the stationary points are

$$(0, 1) \qquad (1, 1) \qquad (2, 1)$$

Differentiating (6.3.8) and (6.3.9)

$$f_{xx}(x, y) = 12x^2 - 24x + 8 = 4(3x^2 - 6x + 2)$$
$$f_{xy}(x, y) = 0 \qquad (6.3.12)$$

and

$$f_{yy} = -6$$

At $(0, 0)$

$$f_{xx} = 8, f_{yy} = -6, f_{xy} = 0 \text{ therefore } \Delta = 8(-6) - 0 = -48 < 0$$

At $(1, 1)$

$$f_{xx} = -4, f_{yy} = -6, f_{xy} = 0 \text{ and } \Delta = 24 > 0$$

At $(2, 1)$

$$f_{xx} = 8, f_{yy} = -6, f_{xy} = 0 \text{ and } \Delta = -48 < 0$$

Hence the points $(0, 0)$ and $(2, 1)$ are saddle points, and the point $(1, 1)$ is a maximum point.

138

(ii) Find the minimum value of the function

$$f(x, y) = xy + 8\left(\frac{1}{x} + \frac{1}{y}\right) \qquad x \neq 0 \qquad \text{and} \qquad y \neq 0$$

The first derivatives are

$$f_x(x, y) = y - \frac{8}{x^2} \qquad \text{and} \qquad f_y(x, y) = x - \frac{8}{y^2} \qquad (6.3.13)$$

so that for a stationary point

$$y - \frac{8}{x^2} = 0 \qquad x - \frac{8}{y^2} = 0 \qquad (6.3.14)$$

Elimination of y between these two equations gives

$$x - 8\frac{x^4}{64} = 0$$

that is

$$8x - x^4 = 0 \qquad (6.3.15)$$

Now $x \neq 0$, therefore (6.3.15) becomes

$$8 - x^3 = 0$$

giving $x = 2$.

Equation (6.3.14) then gives $y = \frac{8}{4} = 2$ and the stationary point is the point $(2, 2)$.

The second derivatives are

$$f_{xx}(x, y) = \frac{16}{x^3} \qquad f_{xy}(x, y) = 1 \qquad f_{yy}(x, y) = \frac{16}{y^3}$$

At the point $(2, 2)$

$$f_{xx} = 2 \qquad f_{xy} = 1 \qquad f_{yy} = 2$$

Therefore at this point $\Delta = 2 \times 2 - 1 = 3 > 0$ and $f_{xx} > 0$ so the point is a minimum.

The minimum value of $f(x, y)$ therefore occurs at the point $(2, 2)$ and is

$$f(2, 2) = 4 + 8(\tfrac{1}{2} + \tfrac{1}{2}) = 12$$

(iii) Discuss the nature of the stationary point at $(0, 0)$ for each of the following functions

$$f(x, y) = x^4 - x^2 y^2 + y^4 \qquad f(x, y) = x^4 - y^4$$

139

In each case it is easily verified that $(0, 0)$ is the only solution of $f_x(x, y) = 0$ and $f_y(x, y) = 0$.

In addition, for both functions

$$f_{xx}(0,0) = f_{xy}(0,0) = f_{yy}(0,0) = 0$$

and an alternative approach is necessary.

Consider the function $f(x, y) = x^4 - x^2y^2 + y^4$. Let (x, y) lie on the line $y = mx$, then

$$f = (m^4 - m^2 + 1)x^4$$
$$= \{(m^2 - \tfrac{1}{2})^2 + \tfrac{3}{4}\}x^4$$

Since $(m^2 - \tfrac{1}{2})^2 + \tfrac{3}{4} > 0$ for all values of m, the function is always positive for $x \neq 0$ and therefore has a minimum at the origin where its value is zero.

Consider the function $f(x, y) = x^4 - y^4$. Let (x, y) lie on the line $y = mx$, then

$$f = (1 - m^4)x^4$$

If $|m| < 1$, then $1 - m^4 > 0$ and f has a minimum at $x = 0$

If $|m| > 1$, then $1 - m^4 < 0$ and f has a maximum at $x = 0$

Thus the behaviour of the function depends on the line of approach, and consequently the origin is a saddle point.

6.4   Functions of More than Two Variables

The method developed in Section 6.3 can be extended to cover functions of many variables. If a function $f(x_1, x_2, \ldots, x_n)$ of n variables has a stationary value at $(\alpha_1, \alpha_2, \ldots, \alpha_n)$ then it is necessarily stationary with respect to each variable in turn, therefore

$$\frac{\partial f}{\partial x_1} = \frac{\partial f}{\partial x_2} = \cdots = \frac{\partial f}{\partial x_n} = 0 \qquad (6.4.1)$$

at the point. These n equations are sufficient to locate the point $(\alpha_1, \alpha_2, \ldots, \alpha_n)$.

If $h_1, h_2, \ldots, h_n$ are small quantities the nature of the stationary point is determined from the increment

$$f(\alpha_1 + h_1, \alpha_2 + h_2, \ldots, \alpha_n + h_n) - f(\alpha_1, \alpha_2, \ldots, \alpha_n) \qquad (6.4.2)$$

In the general case, Taylor's theorem for a function of n variables reduces this expression to the quadratic form

$$\frac{1}{2} \sum_{i=1}^{n} \sum_{j=1}^{n} h_i h_j \frac{\partial^2 f}{\partial x_i \partial x_j} \qquad (6.4.3)$$

140

and the nature of the stationary point depends on the sign of this quadratic form.

For a function of two variables it was seen that the criteria for maximum or minimum could be reduced to two expressions involving the second derivatives. In the case of a function of n variables, n combinations of the second derivatives are required.

In particular for a function $f(x, y, z)$ of three variables the criteria can be given as follows.

Let

$$D_1 = \begin{vmatrix} f_{xx} & f_{xy} \\ f_{xy} & f_{yy} \end{vmatrix} = f_{xx}f_{yy} - f_{xy}^2 \tag{6.4.4}$$

and

$$D_2 = \begin{vmatrix} f_{xx} & f_{xy} & f_{xz} \\ f_{xy} & f_{yy} & f_{yz} \\ f_{xz} & f_{yz} & f_{zz} \end{vmatrix}.$$
$$= f_{xx}f_{yy}f_{zz} + 2f_{yz}f_{zx}f_{xy} - f_{xx}f_{yz}^2 - f_{yy}f_{zx}^2 - f_{zz}f_{xy}^2 \tag{6.4.5}$$

then

if $f_{xx} > 0$, $D_1 > 0$ and $D_2 > 0$, the stationary point is a minimum;
if $f_{xx} < 0$, $D_1 > 0$ and $D_2 < 0$, the stationary point is a maximum.

6.5 Constrained Maxima and Minima—Lagrange Multipliers

In the previous sections, the discussion of extreme values was confined to those situations where all the variables are independent. This is not always the case and it may be important to find stationary values when the points are constrained to lie on a given surface.

The determination of the shortest distance from the origin of coordinates to the surface $g(x, y, z) = 0$ is a typical example. Here the distance d is given by

$$d^2 = x^2 + y^2 + z^2 \tag{6.5.1}$$

and the problem is to minimise that function subject to the constraint

$$g(x, y, z) = 0 \tag{6.5.2}$$

For some functions g it is possible to express z explicitly as a function of x and y, then express d^2 as a function of the two

141

variables x, y and apply the techniques already available for functions of two independent variables.

When $g(x, y, z)$ is more complicated, z may only be determined implicitly and an alternative technique is required.

The following method serves to locate stationary points but does not determine their nature.

The problem is to find the stationary points of the function $F = f(x, y, z)$ subject to the constraint $g(x, y, z) = 0$. The stationary points are determined as a solution of the equations

$$\left(\frac{\partial F}{\partial x}\right)_y = 0 \qquad \left(\frac{\partial F}{\partial y}\right)_x = 0 \tag{6.5.3}$$

Since z is a function of x and y, given by $g(x, y, z) = 0$, these equations can be written as

$$0 = \left(\frac{\partial F}{\partial x}\right)_y = \frac{\partial f}{\partial x} + \frac{\partial f}{\partial z} \cdot \frac{\partial z}{\partial x} \tag{6.5.4}$$

$$0 = \left(\frac{\partial F}{\partial y}\right)_x = \frac{\partial f}{\partial y} + \frac{\partial f}{\partial z} \cdot \frac{\partial z}{\partial y} \tag{6.5.5}$$

The implicit relationship $g(x, y, z) = 0$ can now be differentiated with respect to x and y to give

$$\frac{\partial g}{\partial x} + \frac{\partial g}{\partial z} \cdot \frac{\partial z}{\partial x} = 0 \tag{6.5.6}$$

$$\frac{\partial g}{\partial y} + \frac{\partial g}{\partial z} \cdot \frac{\partial z}{\partial y} = 0 \tag{6.5.7}$$

On elimination of $\partial z/\partial x$ and $\partial z/\partial y$ these four equations give

$$\frac{\partial f}{\partial x} + \lambda \frac{\partial g}{\partial x} = 0 \tag{6.5.8}$$

$$\frac{\partial f}{\partial y} + \lambda \frac{\partial g}{\partial y} = 0 \tag{6.5.9}$$

where λ is a constant satisfying

$$\frac{\partial f}{\partial z} + \lambda \frac{\partial g}{\partial z} = 0 \tag{6.5.10}$$

Together with the original constraint this system of three equations is sufficient to locate the stationary point (x, y, z) and the constant λ.

The constant λ is called a Lagrange multiplier.

An equivalent approach is to construct the function

$$\phi(x, y, z) = f(x, y, z) + \lambda g(x, y, z) \qquad (6.5.11)$$

where λ is a constant to be determined. The stationary points are then given by solving the equations

$$\frac{\partial \phi}{\partial x} = \frac{\partial \phi}{\partial y} = \frac{\partial \phi}{\partial z} = 0 \qquad (6.5.12)$$

and the constraint

$$g(x, y, z) = 0 \qquad (6.5.13)$$

The problem can be generalized to find stationary points for a function $f(x_1, x_2, \ldots, x_n)$ of n variables subject to $k(<n)$, constraints $g_i(x_1, x_2, \ldots, x_n) = 0, i = 1, 2, 3, \ldots, k$.

Let

$$\phi(x_1, x_2, \ldots, x_n) = f(x_1, x_2, x_3, \ldots, x_n) + \sum_{i=1}^{k} \lambda_i g_i(x_1, x_2, \ldots, x_n)$$

$$(6.5.14)$$

where the λ_i are undetermined Lagrange multipliers. The stationary points and the λ_is are given by the $n + k$ equations

$$\frac{\partial \phi}{\partial x_j} = 0 \qquad j = 1, 2, \ldots, n$$

$$g_i(x_1, x_2, \ldots, x_n) = 0 \qquad i = 1, 2, \ldots, k \qquad (6.5.15)$$

Examples

(i) Determine the stationary points of the function

$$f(x, y, z) = x^2 + y^2 + z^2$$

subject to the condition $xyz = 1$.

Let $\phi = x^2 + y^2 + z^2 + \lambda(xyz - 1)$

then

$$\frac{\partial \phi}{\partial x} = 2x + \lambda yz \qquad \frac{\partial \phi}{\partial y} = 2y + \lambda xz \qquad \frac{\partial \phi}{\partial z} = 2z + \lambda xy$$

Therefore the stationary points are given by

$$2x + \lambda yz = 0 \qquad 2y + \lambda zx = 0 \qquad 2z + \lambda xy = 0$$

and
$$xyz = 1$$

Hence, solving these four equations the stationary points are

$$(1, 1, 1) \qquad (-1, -1, 1) \qquad (-1, 1, -1) \qquad (1, -1, -1)$$

(ii) Find the stationary values of $f(x, y, z) = xy + yz + zx$ subject to the condition $x^2 + y^2 + z^2 = 1$

Let
$$\phi = xy + yz + zx + \lambda(x^2 + y^2 + z^2 - 1)$$

then
$$\frac{\partial\phi}{\partial x} = y + z + 2\lambda x \qquad \frac{\partial\phi}{\partial y} = z + x + 2\lambda y$$

and
$$\frac{\partial\phi}{\partial z} = x + y + 2\lambda z$$

Thus the stationary points are given by the equations

$$2\lambda x + y + z = 0$$
$$x + 2\lambda y + z = 0$$
$$x + y + 2\lambda z = 0$$

and
$$x^2 + y^2 + z^2 = 1$$

For a non-zero solution of the first three equations

$$0 = \begin{vmatrix} 2\lambda & 1 & 1 \\ 1 & 2\lambda & 1 \\ 1 & 1 & 2\lambda \end{vmatrix} = 2(2\lambda - 1)^2(\lambda + 1)$$

Therefore, $\lambda = -1$ or $\lambda = \frac{1}{2}$.
When $\lambda = -1$ the equations are

$$-2x + y + z = 0$$
$$x - 2y + z = 0$$
$$x + y - 2z = 0$$

giving $x = y = z$.

144

The remaining equation is $x^2 + y^2 + z^2 = 1$ which then gives $3x^2 = 1$, that is $x^2 = 1/3$. Therefore $x = \pm 1/\sqrt{3}$.

Hence the two stationary points are

$$(1/\sqrt{3}, 1/\sqrt{3}, 1/\sqrt{3}) \qquad (-1/\sqrt{3}, -1/\sqrt{3}, -1/\sqrt{3})$$

At both of these points $f(x, y, z) = 1$.

When $\lambda = \frac{1}{2}$, the first three equations all become

$$x + y + z = 0$$

This equation defines a plane through the origin and consequently all points on the circle of intersection of this plane and the sphere $x^2 + y^2 + z^2 = 1$ are stationary points for the given function $f(x, y, z) = xy + yz + zx$.

The value of $f(x, y, z)$ at each of these points is

$$\begin{aligned}
f(x, y, z) &= xy + yz + zx \\
&= \tfrac{1}{2}\{(x + y + z)^2 - (x^2 + y^2 + z^2)\} \\
&= \tfrac{1}{2}\{0^2 - 1\} = -\tfrac{1}{2}
\end{aligned}$$

Therefore the maximum and minimum values of $xy + yz + zx$ subject to $x^2 + y^2 + z^2 = 1$ are 1 and $-\frac{1}{2}$, respectively.

Exercises

6.1 Find the stationary points of the following functions and determine their nature
  (i) $x^4 - 4x^3 + 4x^2 + 1$
  (ii) $3x^5 - 5x^3 + 2$
  (iii) $x^3/(x + 1)$
  (iv) $\sin 2x - 2\sin x + 2x \qquad -\pi/2 \leqslant x \leqslant \pi/2$
  (v) $x^{-1}\ln x$

6.2 If

$$f(x) = \frac{1 + \cos x}{2 + \sin x},$$

show that the values of $f(x)$ lie between 0 and 4/3.

6.3 Find the maxima, minima and points of inflexion of

$$y = \frac{9x}{(x - 1)^2}$$

6.4 The vertices of a rectangle $ABCD$ of sides x and y lie on a circle of radius R. Show that if the area of the rectangle is a minimum then $x = y = R\sqrt{2}$.

6.5 Show that the graph of the function

$$\frac{x^3}{1 + x^2}$$

has three points of inflexion.

6.6 Find the stationary points of the following functions and determine their nature
(i) $5x^2 + 4xy + 8y^2 - 6x - 12y$
(ii) $x^2 - 4xy + y^2 + 3x + 3y + 4$

(iii) $xy + \dfrac{1}{8x^2} + \dfrac{2}{y^2}$

(iv) $xy \exp -\tfrac{1}{2}(x^2 + y^2)$

6.7 If $f(x, y) = x^2 - 3y^4 + 2y^6$, show that $f_x = f_y = 0$ at $(0,0)$ and at two other points. Show that $f(x, y)$ has a saddle point at $(0,0)$ and minima at the other two points.

6.8 Find the greatest and least distances from the origin to the curve

$$5x^2 - 6xy + 5y^2 = 8$$

6.9 Find the least value of $x^3 + y^3 + z^3$ given that the point (x, y, z) lies on the plane $x + y + z = 1$.

6.10 Find the stationary points of the function $f(x, y, z) = xyz$ subject to the condition $x + y + z = 1$.

6.11 Show that if $x + y + z = 0$ and $x^2 + y^2 + z^2 = 1$ then the magnitude of $1/x + 1/y + 1/z$ is never less than $(27/2)^{1/2}$.

6.12 Show that the function

$$f(x, y, z) = 3x^2 + 5y^2 + 3z^2 - 2xy - 2yz + 2zx - 4x \\ + 6y - 8z + 6$$

has a true minimum at the point $(\tfrac{1}{6}, -\tfrac{1}{3}, \tfrac{7}{6})$.

Answers to Exercises

1.1 (i) $x \geq -1$ (ii) All x (iii) $0 \leq x \leq 2$ (iv) $x > 0$
    (v) $-\sqrt{2} < x < \sqrt{2}$ (vi) $x = 0$ (vii) $x \geq 0$
    (viii) $x \geq 0$

1.2 (i) even   (ii) odd   (iii) even   (iv) even
    (v) even   (vi) odd   (vii) even   (viii) even

1.4 $x = 2n\pi$ $x = (2n + 1)\pi/2$ $x = 2n\pi/5$
    $n = 0, \pm 1, \pm 2, \pm 3, \ldots$

1.7 $\tanh^{-1} A - \tanh^{-1} B = \tanh^{-1}\left(\dfrac{A - B}{1 - AB}\right)$

1.8 (i) $\frac{1}{2}$ (ii) 0   (iii) $\frac{1}{4}$ (iv) 0   (v) 2

1.9 $f(x) \to -\infty$ as $x \to 0^{+}$ $f(x) \to +\infty$ as $x \to 0^{-}$

1.10 (i) $f(x)$ oscillates between 0 and 1
    (ii) $f(x) \to 0$

2.1 (i) $3x^2 + 2x - 5$ (ii) $3 - \dfrac{2}{x^3}$ (iii) $-\dfrac{2x}{(1 + x^2)^2}$

    (iv) $\dfrac{2 - 2x - x^2}{(x^2 - x + 1)^2}$ (v) $\dfrac{2 - x^3}{(1 + x^3)^{3/2}}$

2.2 (i) $-5 \sin 5x$ (ii) $\dfrac{2 \sin x}{(1 + \cos x)^2}$

    (iii) $\sin x \cos^2 x(2 - 5 \sin^2 x)$

    (iv) $-\dfrac{(2 + x^2)}{(\sin x + x \cos x)^2}$ (v) $\dfrac{16 \cos x}{(5 + 3 \sin x)^2}$

2.3 (i) $-\dfrac{1}{x^2} \sec^2\left(\dfrac{1}{x}\right)$ (ii) $\dfrac{4}{5 + 3 \sin x}$

    (iii) $-3[1 + \cos(1 + x)]^2 \sin(1 + x)$

    (iv) $2x \sec(1 + x^2)\tan(1 + x^2)$ (v) $\dfrac{2x}{1 + x^4}$

2.4 (i) $2x(\sin 2x + x\cos 2x)$ (ii) $\dfrac{\cos x - (1-x)\sin x}{(1-x)^2}$

(iii) $(\cos x - x\sin x)e^{x\sin x}$

2.5 (i) $4\sinh(4x+1)$ (ii) $2(x+1)\sinh(x^2+2x)$
(iii) $2x\,\text{sech}^2 x^2$
(iv) $\text{sech}\,2x[\text{sech}^2 x - 2\tanh x \cdot \tanh 2x]$

(v) $-\dfrac{2}{(\cosh x - \sinh x)^2}$

2.6 $2^x \ln 2$ $\dfrac{4^{\sqrt{x}}\ln 2}{\sqrt{x}}$ $\{2x\ln(\sin x) + x^2\cot x\}(\sin x)^{x^2}$

$\dfrac{(\ln x)^{\ln x}\{1 + \ln(\ln x)\}}{x}$ $(x^2+1)^x\ln(x^2+1) + 2x^2(x+1)^{x-1}$

2.7 $\dfrac{dy}{dx} = \dfrac{1+t^2}{2t}$

2.8 $-\dfrac{e^x}{(1+e^x)^3}$

2.9 $-\dfrac{b}{a}\cot\theta$

2.10 $\{(x+n)^2 - n\}e^x$

2.12 (i) 0 (ii) $-\frac{1}{2}$ (iii) 0 (iv) 0

3.1 (i) $\dfrac{x^4 + 6x^2 - 3}{3x}$ (ii) $\frac{1}{2}x(x+12) - \frac{8}{3}\sqrt{x(x+3)} + \ln x$

(iii) $\frac{1}{4}x^4 + \frac{5}{3}x^3 + \frac{5}{2}x^2 - 2x$

3.2 (i) $\dfrac{1}{4}\left\{\ln\dfrac{(1+x)^2}{1+x^2} + 2\tan^{-1}x\right\}$ (ii) $\dfrac{1}{3}\ln\left(\dfrac{x-2}{x+1}\right)$

(iii) $\ln[x(x-1)] - \dfrac{1}{x}$ (iv) $\dfrac{1}{4}\left[\ln\dfrac{(1+x)^2}{1+x^2} - \dfrac{2}{1+x}\right]$

3.3 (i) $x + \ln(\sin x - \cos x)$ (ii) $\frac{1}{4}x^2(2\ln x - 1)$
(iii) $x[\ln(1+x^2) - 2] + 2\tan^{-1}x$

(iv) $\dfrac{e^{4x}}{25}(4\sin 3x - 3\cos 3x)$

3.4 (i) $-2(3 + \cos\theta)^{1/2}$ (ii) $7\theta - 14\cos\theta - 3\sin 2\theta + \frac{8}{3}\cos^3\theta$

3.5 (i) $x + \ln(\sin x - \cos x)$ (ii) $\ln\left(\tan\frac{1}{2}x\right) + \dfrac{2}{1 + \tan\frac{1}{2}x}$

3.6 (i) $\frac{1}{2}\{\sin^{-1}x - x\sqrt{(1 - x^2)}\}$ (ii) $3\sin^{-1}x - 2\sqrt{(1 - x^2)}$

   (iii) $\dfrac{\sqrt{3}}{2}$ (iv) $\ln\left(\dfrac{3}{2}\right)$

3.7 (i) $\frac{1}{8}\{\tan^{-1}(2\cos t) - 2\cos t\}$ (ii) $t - \tan t + \frac{1}{3}\tan^3 t$

3.8 (i) $\dfrac{1}{20}$ (ii) $-\dfrac{4}{3}$ (iii) $\dfrac{\pi}{81}(\pi^2 - 6)$ (iv) $\dfrac{1}{2}$

3.9 (i) $\dfrac{28\sqrt{2}}{9}$ (ii) $\ln\left(\dfrac{4}{3}\right) - \dfrac{1}{6}$ (iii) $\dfrac{\pi}{2}$

3.10 $\frac{1}{2}(\pi - 2)$

3.11 (i) $\dfrac{3\pi}{16}$ (ii) $\dfrac{3\pi^2 + 16}{64}$ (iii) $\dfrac{3\pi a^4}{16}$

3.12 18
3.13 $\frac{1}{3}a^2\ln 2$

3.14 $\dfrac{\pi}{4} - \dfrac{1}{3}$

3.15 $\frac{8}{15}$

4.1 1
4.2 $f(x, y)$ is discontinuous and non-differentiable at $(0, 0)$.
4.3 (i) $8x - 1, 6y + 1$
   (ii) $\cos x \, e^{(\sin x + \cos y)}, -\sin y \, e^{(\sin x + \cos y)}$
   (iii) $y \sec^2(xy), x \sec^2(xy)$

   (iv) $-\dfrac{y}{x(x^2 - y^2)^{1/2}}, \dfrac{1}{(x^2 - y^2)^{1/2}}$

   (v) $y \tan(xy), x \tan(xy)$
4.4 $(1 + xy)e^{xy}$
4.5 -5 7
4.6 $2\sin y - y^2 \sin x$ $2x\cos y + 2y\cos x$
   $-x^2\sin y + 2\sin x$

4.10 (i) $2u$ (ii) $2u$ (iii) $\dfrac{u}{2(u^2 + v^2)}$ (iv) $\dfrac{2(u^2 + v^2)}{u}$

149

4.13 $\mathrm{d}U = \dfrac{R(3pv^2 - a)}{2(pv^2 - a)}\,\mathrm{d}T + \dfrac{av}{(a - pv^2)}\,\mathrm{d}p$ $\qquad \left(\dfrac{\partial U}{\partial p}\right)_T = \dfrac{av}{(a - pv^2)}$

$\left(\dfrac{\partial U}{\partial T}\right)_p = \dfrac{R(3pv^2 - a)}{2(pv^2 - a)}$

4.15 $\dfrac{\mathrm{e}^{-2x}}{(x + 1)^2 + y^2}$ $\qquad \mathrm{e}^{2x}(x + 1)^2 + y^2$

4.18 $\dfrac{2V^2}{g}\cos 2\alpha \cdot \Delta\alpha$ $\qquad \dfrac{2V}{g}\cos \alpha \cdot \Delta\alpha$

4.19 1%

4.20 8.8%

5.1 (i) $1 - \frac{3}{2}x^2 + \frac{7}{8}x^4 + \cdots$
   (ii) $1 + \frac{1}{2}x^2 + \frac{5}{24}x^4 + \cdots$
   (iii) $x + \frac{1}{3}x^3 + \frac{2}{15}x^5 + \cdots$

5.2 (i) $\displaystyle\sum_{n=0}^{\infty} \frac{x^{2n}}{(2n)!}$ $\qquad\qquad$ (ii) $\displaystyle\sum_{n=0}^{\infty} \frac{x^{2n+1}}{(2n+1)!}$

   (iii) $\dfrac{1}{2}\displaystyle\sum_{n=1}^{\infty} \frac{(-1)^{n-1}(2x)^{2n}}{(2n)!}$

   (iv) $\displaystyle\sum_{n=0}^{\infty} \frac{(2n)!}{n!n!}\left(\frac{x}{2}\right)^{2n}$ $\qquad$ (v) $\displaystyle\sum_{n=0}^{\infty} \frac{(-1)^n x^{2n+2}}{n+1}$

   (vi) $\displaystyle\sum_{n=0}^{\infty} \frac{(-1)^n x^{2n+1}}{2n+1}$ $\qquad$ (vii) $\displaystyle\sum_{n=0}^{\infty} \frac{(2n)!x^{2n+1}}{(2n+1)n!n!4^n}$

5.3 $(1 - x^2)^{-1/2} = 1 + \dfrac{1}{2}x^2 + \dfrac{3}{8}x^4 + \dfrac{5}{16}x^6 + \cdots$

   $\cos^{-1} x = \dfrac{\pi}{2} - x - \dfrac{1}{6}x^3 - \dfrac{3}{40}x^5 - \dfrac{5}{112}x^7 - \cdots$

5.5 $\tanh x = x - \frac{1}{3}x^3 + \cdots$

5.6 $1 + x^2 - 3x^3 + x^4 - \cdots$

   $\displaystyle\sum_{n=0}^{\infty} x^{2n} + \sum_{n=0}^{\infty} [(-1)^n 2^n - 1]x^{2n+1}$

5.7 (i) $1 - (x + y) + (x + y)^2 + \cdots$
   (ii) $\mathrm{e}\{1 + x + y + \frac{1}{2}(x^2 + 4xy + y^2) + \cdots\}$
   (iii) $2y - 2xy + \frac{2}{3}y(3x^2 + y^2) + \cdots$

150

6.1 (i) Minima at $x = 0$ and $x = 2$, maximum at $x = 1$
     (ii) Maximum at $x = -1$, point of inflexion at $x = 0$, minimum at $x = 1$
     (iii) Minimum at $x = -\frac{3}{2}$, point of inflexion at $x = 0$
     (iv) Maxima at $x = -\frac{\pi}{2}$ and $x = \frac{\pi}{3}$, minima at $x = -\frac{\pi}{3}$ and $x = \frac{\pi}{2}$
     (v) Maximum at $x = \mathrm{e}$

6.3 Minimum at $x = -1$, point of inflexion at $x = -2$

6.5 Points of inflexion at $x = -3$, $x = 0$ and $x = 3$

6.6 (i) Minimum at $\left(\frac{1}{3}, \frac{2}{3}\right)$ (ii) saddle point at $\left(\frac{3}{2}, \frac{3}{2}\right)$
     (iii) Minima at $\left(-\frac{1}{2}, -2\right)$ and $\left(\frac{1}{2}, 2\right)$
     (iv) Maxima at $(-1, -1)$ and $(1, 1)$, minima at $(-1, 1)$ and $(1, -1)$, saddle point at $(0, 0)$

6.7 Minima at $(0, -1)$ and $(0, 1)$

6.8 Greatest distance 2, least distance 1

6.9 $\frac{1}{9}$ at point $\left(\frac{1}{3}, \frac{1}{3}, \frac{1}{3}\right)$

6.10 $(0, 0, 1)$, $(0, 1, 0)$, $(1, 0, 0)$ and $\left(\frac{1}{3}, \frac{1}{3}, \frac{1}{3}\right)$

Index

153

p.87–
learn ✓
p.103